Citizens in the 'Smart City'

This book critically examines 'smart city' discourse in terms of governance initiatives, citizen participation, and policies which place emphasis on the 'citizen' as an active recipient and co-producer of technological solutions to urban problems.

The current hype around 'smart cities' and digital technologies has sparked debates in the fields of citizenship, urban studies, and planning surrounding the rights and ethics of participation. It has also sparked debates around the forms of governance these technologies actively foster. This book presents new sociotechnological systems of governance that monitor citizen power, trust-building strategies, and social capital. It calls for new data economics and digital rights for a city founded on normative ideals rather than neoliberal ones. It adopts a prescriptive approach, arguing that a 'reloaded' smart city should foster citizenship as a new set of civil and social rights and the 'citizen' as a subject vested with active and meaningful forms of participation and political power. Ultimately, the book questions the utility of the 'smart city' project for radical municipalism, proposing a technological enough but more democratic city – an 'intelligent city', in fact.

Offering useful contributions to 'smart city' initiatives for the protection of emerging digital citizenship rights and socially accrued benefits, this book will draw the interest of researchers, policymakers, and professionals in the fields of urban studies, urban planning, urban geography, computing and technology studies, urban politics, and urban economics.

Paolo Cardullo is Senior Researcher at IN3 (Internet Interdisciplinary Institute), Universitat Oberta de Catalunya in Barcelona, where he was awarded the Beatriu de Pinós scholarship in 2019. His research interests are in urban geography, smart cities, and urban governance. His previously published titles with Routledge are *Gentrification in the Mesh?* (2017) and *Sniffing the City* (2014).

Routledge Studies in Urbanism and the City

For more information about this series, please visit www.routledge.com/
Routledge-Studies-in-Urbanism-and-the-City/book-series/RSUC

Citizens in the 'Smart City'
Participation, Co-production, Governance

Paolo Cardullo

Routledge
Taylor & Francis Group

LONDON AND NEW YORK

First published 2021
by Routledge
2 Park Square, Milton Park, Abingdon, Oxon OX14 4RN

and by Routledge
605 Third Avenue, New York, NY 10017

First issued in paperback 2022

Routledge is an imprint of the Taylor & Francis Group, an informa business

© 2021 Paolo Cardullo

The rights of Paolo Cardullo to be identified as the author of this work has been asserted by him in accordance with sections 77 and 78 of the Copyright, Designs and Patents Act 1988.

All rights reserved. No part of this book may be reprinted or reproduced or utilised in any form or by any electronic, mechanical, or other means, now known or hereafter invented, including photocopying and recording, or in any information storage or retrieval system, without permission in writing from the publishers.

Trademark notice: Product or corporate names may be trademarks or registered trademarks, and are used only for identification and explanation without intent to infringe.

Publisher's Note
The publisher has gone to great lengths to ensure the quality of this reprint but points out that some imperfections in the original copies may be apparent.

British Library Cataloguing-in-Publication Data
A catalogue record for this book is available from the British Library

Library of Congress Cataloging-in-Publication Data
A catalog record for this book has been requested

ISBN 13: 978-0-367-55995-3 (pbk)
ISBN 13: 978-1-138-34394-8 (hbk)
ISBN 13: 978-0-429-43880-6 (ebk)

DOI: 10.4324/9780429438806

Typeset in Times New Roman
by Apex CoVantage, LLC

"Sorry, Dad, I did not finish my book on time for you to read it . . ."

Contents

Tables

Acknowledgements

The book was conceived during my first postdoctoral post at Maynooth University working on The Programmable City project funded by a European Research Council Advanced Investigator award (ERC-2012-AdG-323636-SOFTCITY). Its scholarly framework and empirical research are mostly due to working closely with Professor Rob Kitchin and his team of researchers. Other research has been undertaken soon after my PhD at Goldsmiths, University of London; during my second postdoctoral post at the Business School of Maynooth University; and finally, as incipient researcher at IN3 research centre of the Universitat Oberta de Catalunya, where I recently started as Beatriu de Pinós scholar. Editing the manuscript final draft during the 'lockdown' has not been an easy task and I would like to thank very much my beloved family – Luca, Anna, Emidio, and Edy – for their patience and support.

Acknowledgements

Introduction

The age of the 'Smart City' is upon us!

It sounds almost like a truism these days that cities around the globe are or should be getting 'smarter'. The message is clear, but it is also repeated at almost every meeting, presentation, or gathering of city officials, high-tech experts, and academics: The world is changing fast on the terms of algorithm-led processing and cities need to catch up with this flow or perish.

Mundane actions are increasingly directed, dictated, or modulated by partly invisible or non-optional processes of data gathering, analysis, and display which span almost every aspect of our lives. A sense of urgency is often implicit through the mediatic realm – increasingly via Twitter and LinkedIn personal or professional accounts – for an incumbent future where technological solutions promise to fix nearly everything. This is particularly relevant for cities, where most people live, work, and operate on a daily basis: congestion avoiding apps, remotely controlled home assistance, customised shopping experiences, flexible and just-on-time travelling arrangements, ad-hoc services in virtually every aspect of urban life, reduction of pollution, monitoring of body performance, real-time industry competitiveness, budget saving apps, metering of service provision, curtailing of CO^2, and even predictive and voice/face recognition policing.

It is also true that, somehow, cities should prepare themselves to steer, manage, and even try to control these changing infrastructures and the flow of data they generate. Perhaps cash-strapped city councils can rely on efficient private companies for the delivery of essential services via cost cutting technologies? Or perhaps cities can benefit from the availability of data, algorithmic decision-making, and citizen-connected publics, in order to foster meaningful participation? Or perhaps cities want to rethink their priorities, and even go beyond the 'smart city', in order to reduce social inequalities in the distribution of opportunities, housing, and access to critical infrastructures? These three rhetorical propositions offer a modulation of responses, adaptations, and different governance arrangements cities are putting forward with regard to the 'smart city'. In the first scenario, cities are run like a private tech company or are run by large tech companies (Kitchin, Graham, Mattern, & Shaw, 2019). The second one draws on a range of opportunities the emerging data infrastructure seems to offer to cities if they are able or willing to facilitate such a technological change (Morozov & Bria, 2018). The final proposition, instead, suggests cities can re-appropriate the political debate on what is a more just and equal city, steering or changing socioeconomic goals towards the 'right to the smart city' (Cardullo, Kitchin, & Di Feliciantonio, 2019), or towards something different – an 'intelligent city'.

Clearly, this is where the agreement ends. The way in which urban issues are posed and dealt with, the modalities through which technologies are delivered and operate, and the responses and levels of participation citizens are called to act on can vary from city to city, from one technological solution to the other, and even from the implementation stage of the same technology to the next one. In other words, the 'smart city' is never a unique packaged solution, 'one size fits all', but rather an ensemble (or assemblage) of myriad of projects, voices, decisions, reports, initiatives, proposals, pilots, roles, responsibilities, deliberations, and executions. Thus, this book argues that the 'smart city' is, first and foremost, a cultural construction: a matter of political choices and consensus, which set the priorities, and of meaningful participation and democratic urban governance, which set the context. In other words, the 'smart city' should be understood as an intrinsically *political process*, rather than a purely technical one – a process that belongs to, and ultimately calls into question, the very democratic functioning of urban societies in the third millennium.

This book is then partly about why smart technologies – ways of collecting, storing, displaying and circulating data, services, and functions, often remotely via the Internet – became the 'smart city': an almost unique framework of prepackaged solutions led by market principles and neoliberal policies which works towards an already prescribed urban future and in which the 'citizen' occupies a very instrumental and mostly passive role. The first part of this book asks, more in detail: What kind of cultural imaginary and political economy ideals are deployed in the notion of the 'smart city'? How are ordinary citizens framed and operationalised within it? The second part of the book develops a more normative argument around the 'right to the smart city', asking: What forms of social justice and citizen empowerment can an alternative 'smart city' offer? And how can an alternative 'smart city' be imagined and made possible beyond the neoliberal ideals? Further, do we need the 'smart city' concept, after all? Can we imagine organising city priorities and conducting urban life *with* technologies, rather than dictated *by* technologies?

These technologies – environmental sensors, app-led urban mobility, Internet of Things devices, traffic control systems, data dashboards, personal voice assistants, and so on – are, however, increasingly deployed in every realm of urban and personal living. The diversity of 'smart city' initiatives is illustrated in Table 0.1, which orders these initiatives through the different domains.

Table 0.1 makes it clear how smart technologies are becoming pervasive in every aspect of the life of cities (civic, economic, and social domains). From the very functioning of city governance and security to the intimate spaces of home and the body, sensors and related software for extraction, transmission, and eventual analysis of data are implemented or used by a vast number of

Table 0.1 Different forms of 'smart city' technologies

Domain	Example technologies
Government	E-government systems; online transactions; city operating systems; performance management systems; urban dashboards
Security and emergency services	Centralised control rooms; digital surveillance; predictive policing; coordinated emergency response
Transport	Intelligent transport systems; integrated ticketing; smart travel cards; bikeshare; real-time passenger information; smart parking; logistics management; transport apps; dynamic road signs; mobility apps; ride-share services
Energy	Smart grids; smart meters; energy usage apps; smart lighting
Waste	Compactor bins and dynamic routing/collection
Environment	IoT sensor networks (e.g., pollution, noise, weather; land movement; flood management); dynamically responsive interventions (e.g., automated flood defences)
Buildings	Building management systems; sensor networks
Homes	Smart meters; app-controlled smart appliances; digital personal assistants

Source: Kitchin (2016a)

city dwellers, organisations, service providers, and city executives. These technologies are increasingly deployed in, and woven into the infrastructural fabric of, cities where a critical mass of users (more often, customers) is located and operates.

The massive introduction and deployment of mostly networked digital devices have been relentless in the last 10–15 years and it is set to rise. High-tech giant Dell, for instance, estimates 200 billion connected devices by 2031.[1] While traditionally the Internet has facilitated interaction between humans, the current socio-technological landscape is starting to be dominated by a different configuration. Recording, storing, or transmission are the *modus operandi* of smart technologies, their reasons to be: they are now known as the 'Internet of Things' (IoT). These 'things' are digital devices that typically have a communication interface, processing and storage units, and sensors for detection of environmental changes or for service provision to other clients. The combination of short-range mesh networks and the wider cellular network can provide wireless connectivity to these 'things' in order to exchange data over the wider Internet. On the one hand, people in the reach of digital technologies take overlapping roles while using, testing, or simply being counted by them, potentially showing a great deal of complexity as dealers of different data points operating simultaneously on different scales (the body, the neighbourhood, the urban, and the global). On the other hand, people's daily practices are in the reach of digital devices and applications, virtual augmentation and communication, on-time and ad-hoc occurrences, and real-time monitoring of systems that inherently surveil and control.

Ultimately, what makes the 'smart city' inherently urban are the coming together of two trends. On the one hand, the recipients of these services and functions are intimately linked to the urban *milieux* where they live, work, and exchange. On the other hand, cities might include some of the above services and initiatives under the label of the 'smart city', for various reasons: for instance, in order to respond to the competing demands put on their austerity-eroded budgets by lobbying groups and transnational coalitions. It follows that the 'smart city' is never a unique system or one specific modality of operations involving algorithm-operated technologies. Thus, a working definition of 'smart city' can focus on the assemblage of various sociotechnological initiatives, projects, and pilots that seek to predict and control city living by way of remote, interconnected, and technology-led forms of governance. The incredible variety of 'smart city' solutions, which results from this assembling of functions and roles, can best be appreciated in specific case studies, such as those presented in the second part of the book. Here, instead, I want to focus on their rather consistent characteristics, which are summed up by the following five 'edge points'.

Edge points

First, smart technologies are always somehow linked to an algorithm-led response, even only for a card payment or for an instance of communication. Ultimately, algorithmic functions represent the 'smart' bit in the 'smart city'. In this regard, it is worthy to remember that the adjective 'smart' has different meanings in the 'smart city' debate, and these meanings relate greatly to the different cultural constructions and perspectives that sustain such a debate. For instance, 'smart' might mean 'clever' and this is, in my opinion, the most commonly used meaning, in the sense that the device, sensor, or app is able to detect or even predict change – for example, in the surrounding environment – without human or third-party input. Some of these responses are automatic and autonomous (second-cybernetic order of functioning) for there is no apparent need of user's input or agentive instruction, such as in traffic light control systems or location tracking. Another oft-used meaning of 'smart' is 'elegant' or 'cool', which also attaches well to the increasingly sophisticated design of devices, likely to become status symbols in the relevant community of practice (e.g., new smart phones or watches), or with regard to places of habitation (e.g., smart home and voice-operated personal assistants). Finally, 'smart' can mean 'intelligent', and this is probably the meaning most used by activists trying to shift the 'smart city' away from the 'clever' and 'cool' positions, made of automatic and exclusive solutions, to a more inclusive, just, and environmentally sound city (e.g., city services delivered via local cooperatives of workers, or community-led processes of monitoring noise and pollution). The ethos of this book is exactly in questioning uniqueness of meanings for the 'smart city' while proposing a critical reading of its predominant modality: the algorithm-led process of understanding and operating the city thought real-time analytics of swaths of data from different smart technologies.

Second, a consistent characteristic of smart technologies is that they always involve some transmission of information or data from one agent (machinic, human-operated, or human) to another. On a more superficial level, transmission entails a degree of communication and feedback that would otherwise be impossible, typically in 'Voice Over IP' (VOIP) calls across the globe or in forms of participation over city platforms and private apps, when these are available. The benefits of increased communication between people are arguably the most popular aspects of smart phones, computers, social media, and the Internet as global platforms. On a deeper level though, communication and exchange happen in milliseconds also between algorithms that regulate other aspects of daily life such as financial interfaces and banking transactions, or IoT devices and servers. Moreover, services to production function increasingly within 'just-on-time' processes, which tend to pre-empt human-oriented responses (e.g., they can predict supply and determine orders before demand arises within the production chain). This is very relevant to the scope of this book, in the sense that citizen participation is often limited or constrained within already prescribed software environments (Gabrys, 2014). Processes and feedback that are prevalently automatic and autonomous, for instance, are beyond the agentive reach of their publics, bolstering data extraction from passive data-points for corporate and state actors, with users becoming themselves products of data retrieval and treatment, and data circulation practices. Understanding the work that algorithms do in maintaining an order of actions and responses, or failing to do so, can thus be crucial to other aspects of the 'smart city' debate. While this book does not look at algorithms *per se* (e.g. Kitchin, 2014a, 2017; O'Neil, 2017), there is a discussion about algorithmic rationality with regards to governance and to the roles citizens effectively perform when relating to such technological environments.

Third, smart technologies need some form of connectivity or infrastructural support for data exchange (Kitchin & Dodge, 2019). While the underlying infrastructure, or 'backbone', is almost always taken for granted, it is important to address the physical infrastructures that deliver the Internet in cities – its bandwidth, its resilience, and especially its fair access and distribution. To the ideal of interconnected citizens, I dedicate a study of community broadband provision in London, comparing this with two business models of public intervention in the United States: the case studies of Chattanooga, Tennessee, and New York City. For me, equal access to unlimited and unmonitored Internet can be thought of in terms of collective human rights and ought to increase fairness and justice in the 'smart city' of the future, whether in the Global North or South. Instead, the battle for speed and reliability of the Internet connectivity is currently fostering exclusivity (super connected areas or hubs) and splintered provision (Graham & Marvin, 2001). In their pioneering study of Internet mapping, Dodge and Kitchin (2001, p. 10) maintain that "accessing cyberspace is fragmented along traditional spatial and social divisions, with infrastructure density and variety being closely related to areas of wealth." Inequality of access has become a striking paradox at the heart of the 'smart city': how can this super connected and increasingly complex urban environment work without a strong backbone or without a capillary distribution to as many users/clients/customers/citizens as possible? One bottle-neck for adoption of the 'smart city' would be, paradoxically, exactly this scarcity, which capitalism is so good at creating via market imperatives of privatization and efficiency. I develop an argument for a public-led provision of the Internet – at the time of writing, the most common way of doing this seems to be 'Broadband over Fibre' – which operates through the direct involvement of users and communities of practice. This is not to endorse a solution to the 'smart city' adoption issue. As critical scholars would object (e.g., Couldry, 2018), without changing the present conditions of the Internet as a source of data extraction (that is, how the Internet is governed and controlled by corporations and states), more or better access to it does not equate to more freedom. The argument for a universal public-led provi-sion of the Internet would rather advance a critical approach to the 'smart city' for the People and not for the Things, and for 'the many' rather than 'the few'. In order to reduce digital divides and guarantee a basic form of participation in the 'smart city', we cannot dismiss the issue of a free, or affordable, and equal Internet provision. At the same time, the Internet has to be considered a critical infrastructure, like water or heat (one of the rights that makes the 'right to the smart city' operative, so to speak).

Fourth, smart technologies are pervasive in every aspect of daily life to the point that they purport an ideal of city more akin to the 'social factory 4.0.' The number '4.0' follows the concept of the 'long waves of technological change', as expressed by scholars such as Kondratiev, Shumpeter, and others (e.g., Perez, 2009). The foundations of this idea are not new, drawing on the Italian Autonomist tradition that has looked at the modes of production of advanced capitalism as becoming widespread in society rather than limited to the *loci* of production, such as the sweatshop or the factory. In a nutshell, they maintain that, since labour is now expressed mostly through imagination, creativity, ideas, and soft production – in other words, labour is mostly 'immaterial' (Hardt & Negri, 2009) – capitalism has become 'cognitive', harnessing the collective intelligence of workers and city dwellers beyond the spaces of production. And because this exchange of ideas and creativity happens mostly in urban environments where critical masses of people, users, or creatives, are in tight daily contact, cities are seen as the magnet and the pulsing rhythm of such cognitive labour. While schol-ars have pointed to the processes of communication and networking between people that maintain such an exchange, 'cities 2.0' (Rossi, 2017), the 'smart city' has deepened such an exchange. In my view, immaterial labour as the exchange of creativity and ideas – thus, led by communication and information – does not exhaust the working of the 'social factory', because a large part of smart technologies compels an exchange between machines via algorithmic responses that are *beyond* people's agentive ability and consciousness. As such, the object of such an exchange is not so much, or uniquely, the cognitive process of com-munication; rather, it is predominantly *data* (see Kitchin, 2014b). Through *data*, machines explicate their daily functions and, through *data*, they communicate to other machines or operators behind them, often in milliseconds. This is why the emphasis on the 'smart city' is rightly shifted towards the technological change in the *production* process of data itself, that is, the move from the human-operated communication framework, where human agency is still predominant, to the algorithm-governed social factory. In the first part of the book, I discuss at length this crucial aspect of data extractivism by smart technologies and their roles in everyday life. Here, it is fair to anticipate that the citizen in the 'social factory 4.0' appears to perform predominantly as a 'data point', the means through which data are accrued for all sorts of reasons. The increased importance of data in the very fabric of city living is therefore crucial to understanding the passage to the 'smart city'.

Fifth, the variety and pervasiveness of smart technologies in the everyday life of cities are raising a formidable array of ethical issues in relation to extraction and treatment of data, surveillance and control, as well as privacy, consent, and disclosure of personal and sometimes sensitive data for any purpose: from profiling, social sorting, and anticipatory governance, to marketing of goods and nudging of consumer behaviour. 'Smart city' technologies, the data they generate and the analytics applied to them, can thus have significant negative direct and indirect impacts on peoples' everyday lives. This is because the present configuration of technologies is dependent on systems that inherently surveil and control (Kitchin, 2016b). If this is the case, and the book underlines such a critical perspective on smart technologies, careful political consideration and ethical scrutiny are required at each stage of their design and implementation. At the present, we can count about 70 ethical frameworks, guidelines, and principles on AI, robotics, and data (Floridi, 2019). If, on the one hand, this can be a symptom of how urgent the topic is becoming, on the other hand, it is also a clear sign of general confusion, compliance to existing or forthcoming legislations, and a sign of corporate opportunism. In trying to reduce this 'noise' around the ethics of AI, robots, algorithms, and data handling, in 2018 the European Commission appointed 52 experts to a new High-Level Expert Group on Artificial Intelligence (AI HLEG) who delivered their own framework and guidelines for algorithm-driven and deep learning processes, "Ethics Guidelines for Trustworthy AI." This is designed to guide the AI community in the responsible development and use of "lawful, ethical and robust" AI.

In sum, the 'smart city' is shaping urban space and influencing urbanism, massively. As Kitchin and Dodge (2011) note, urban space is being increasingly translated and represented through algorithmic processes, which in turn determine space and its functioning in differential manners as well as through a modulation of temporal relations. For Kitchin (2019b), the dimensions of code, space, and time are now linked to each other as contiguous and contemporary relations, rather than as sequential causation. In other words, we cannot avoid assessing the working of such technologies in the present while talking of citizenship rights of urban dwellers in a proximate future. At the same time, a discussion on citizenship and rights in advanced technological capitalism cannot bypass a critical evaluation of the role of data and technology deployment and the ethical issues associated with them (for instance, the way in which governance of such citizens in itself is changing thanks to the deployment of dashboards, digital platforms, and mobility apps). It is in this novel space, the space of the 'smart city', that my reflection on the citizen takes place.

To respond to the above critiques, the designers and the deployers of 'smart city' technologies have sought to reframe these as 'citizen-centric', by saying that they privilege the interests, needs, and concerns of citizens. Apart from some experimental advances in citizen science (with sensors and devices deployed for the monitoring of environmental pollution) and some pilots for direct democratic decision making (such as community-led responses to city planning), to date there seems to have been little progress in this regard: this is because such citizen-centric initiatives remain framed within the neoliberal framework of data extraction, efficiency of service delivery, and, essentially, marketisation of city functions (Cardullo & Kitchin, 2019b). In other words, most citizen-centric initiatives only scratch the surface of citizen participation and empowerment, acting as a window-dressing mechanism in a much more vast landscape of tracking-enabled, surveillance-prone, and profit-making technology adoption. The argument starts exactly at the disjuncture between public and private stakeholders' claims around what roles 'smart citizens' have been attributed or perform, and the 'actually existing smart city' (Shelton, Zook, & Wiig, 2015) as a technology of governance for citizen participation or control. Moreover, the current discourse and actual deployment – this is what I call the 'smart city', in inverted commas – is rather instrumental to the neoliberalisation process of cities. Neoliberalisation works towards boasting a policy rationality and forms of governance which we can call with Morozov (2013) "technological solutionism," while fostering aspects of neoliberal citizenship (individualism, self-management, entrepreneurial self, freedom of choice).

Moving from the individual design, trials, and eventual implementation of a particular technology to the 'smart city', as a unique operative concept of scholarship and governmentality, requires a set of conceptual adjustments that needs to be explained. The ways in which the 'smart city' is conceptualised, mobilised, and understood and how this ideal shapes the making of the future city are legitimate research questions. Consequently, the first part of this book debunks the myth of the 'smart city' as currently conceived around a discourse of neoliberal efficiency and technological solutions to urban issues. It shows, on the one hand, how high-tech companies and private service providers have been seeking public funding and steering regulations for new markets preoccupied with sensors, IoT, apps, data centres, connecting devices, safety and surveillance, predictive analytics, mobility, monitoring of homes and bodies, tracking of individual preferences and emotions, and so on. On the other hand, the hype around technology has come to imply outsourcing of government functions to technological processes, which ultimately are cost and labour saving; massive privatisation of service provision for the benefit of the high-tech industry; and procurement practices that encourage efficiency savings over inclusion and access in exchange of personal and operational data for many different purposes. In other words, the 'smart city' has been functioning within an already well-oiled mechanism of neoliberalisation of urban space: according to Hollands (2008), for example, the 'smart city' is the technological expression of this process, which has been in place in cities since the 90s. From the 'entrepreneurial turn' of city policy, where cities have been called to promote themselves and attract foreign investments and capital in the forms of mega events, tourism, and the reception industry, to the 'creative city' through which the urban is refashioned as a cool place to live and work in order to appeal to the emerging creative classes, the hype around city growth has been cast solidly in the imagination of city administrators.

The 'smart city' complements this trend, offering a mix of the above neoliberal policy solutions infused with technological determinism. While setting appropriate goals for cities via systems of urban benchmarking, the neoliberal 'smart city' aims to attract foreign direct investment, offering areas of the city as testbeds to pilot new technologies, fostering innovative indigenous start-up sectors or digital hubs, and attracting mobile creative elites. Thus, intra-city competition stimulates a speculative approach to housing, privatisation of space, and attraction of more affluent buyers, all characteristics of neoliberal urbanism, which conceives urban land via exchange value rather than use value (Kitchin, O'Callaghan, Boyle, Gleeson, & Keaveney, 2012). In other words, the latest hype around smart technologies has reinforced an already winning neoliberal discourse on city growth.

In the second part of the book, I attempt to draw on *alternative* 'smart city' ideas, which are concerned with meaningful participation and forms of citizen power, where technologies are deployed not for the sake of data extraction but for the goals of making urban space more habitable; that is, technologies are thought to be designed and implemented around principles of equality, inclusion, and democratic control of the city by its citizens (or democratic governance arrangements between cities and citizens). It will appear evident soon that such a conceptualisation has to be foregrounded on principles *alternative* to the market-oriented ideals expressed by the neoliberal 'smart city'; rather, on a bundle of rights and entitlements to inhabitation and control of our own urban space: in other terms, on the 'right to the smart city' (Cardullo et al., 2019). For instance, the city of Barcelona has sought to use the existing technology for answering the needs of its citizens rather than adopting new technologies for their own sake. Barcelona has sought to *re-politicize* the smart city and to shift its creation and control away from private interests and the state toward grassroot, civic movements and social innovation. In the end, I will question the utility of the 'smart city' concept for radical urbanism and suggest the need to move beyond this, maybe towards an 'intelligent city'.

Mapping the book content

The first chapter of this book attempts to define the 'smart city', by way of linking this concept to the recent history of urban science. There is a vast literature that unpacks the goals of integrating computerised processes in the life of cities, following the development of urban science and the predominance of quantitative methodologies in making sense of complex, heterogeneous, and unpredictable environments. These theoretical and epistemological discussions are set as a backdrop to the next chapters which ask: How does a discursive construction take place, get traction, and get shared simultaneously across the world? What is the socioeconomic milieu in which the 'smart city' becomes a necessity? We will see how the coming together of a policy framework, driven by neoliberal austerity, and the proliferation of advocacy from industry and academia, which claims to offer easy solutions to 'fix' long-standing urban issues, have been crucial to the success of the 'smart city'.

To this end, Chapter 2 suggests that the socioeconomic rationality of neoliberal urbanism sits comfortably with algorithmic rationality and forms of governance promoted by the 'smart city' discourse. The chapter frames neoliberal urbanism as a set of policies that is rooted in the belief that the market allocates resources in a more efficient manner and, therefore outsources services provision to the private sector; attracts foreign investments and prime workers as policy goal; and promotes city growth in competition with other cities. Neoliberal urbanism and the present configuration of the 'smart city' work together and are deeply intertwined as cultural construction and dominant policy discourse in cities today. While technological solutions to urban problems have been around for some time, the 'smart city' seems to be something a bit different, more integrated with everyday life and implanted in the discourses around the future of cities.

Chapter 3 suggests deploying the concept of the 'post-political' (Swyngedouw, 2011) in order to make sense of decisional processes of implementation and governance of 'smart city' technologies that have little or no political and ethical oversight, while deeply affecting the life of cities and citizens. Proponents of the 'smart city' formulate entrenched 'problems' affecting city life, such as reduction of crime or pollution, in a rather generic way: for instance, levels of CO_2 are always good benchmarks to almost any suggested solution (e.g., electric vehicles), while more concrete goals are avoided – specific mission-oriented objectives such as the reduction in X number of cars on the road – and this can lead to confusion between the overall strategy and the proposed fixes (Mazzucato, 2018b). Whether the 'smart city' idea aims to make city life better using digital technologies for the management and delivery of city services, clearly a lot can be said about how we can get there. The way in which the 'problem' of the city is traditionally framed is a necessary gateway to the suggested solutions: the therapy is prescribed after the symptoms are assessed. However, with the coming together of algorithmic rationality and urban science, technology makes the 'problem' knowable in a certain way, but it *also* determines its solutions.

Chapter 4 addresses 'the citizen' and the variety of theoretical underpinnings we can derive from the scholarly debate on citizenship in the contemporary 'smart city'. Within the neoliberal policy framework for the 'smart city', funding agencies and policy documents have been boasting the formation of a well-organized epistemic community (a knowledge and policy community) and advocacy coalition (a collective of vested interests) operating across scales from global to local, in addition to a cohort of favourably minded technocrats occupying city government functions (Kitchin, Coletta, Evans, Heaphy, & MacDonncha, 2017). This has been usually represented as the Triple Helix: city officials (usually a few members of a newly created 'smart city office'), high-tech industry experts (often from leading brands), and academics (mostly from STEM faculties). The obvious absentee here is 'the citizen' whose behaviour is, however, always called into question and whose needs are already known in advance, and taken care of. Finally, the chapter offers some case studies from Dublin, Ireland, around the 'actually existing' smart citizen, the roles and functions these citizens are called on to occupy in the 'smart city', and the kinds of power relations therefore forged in between algorithms, bits, and Wi-Fi signals.

Chapter 5 offers a panoramic on the Living Labs phenomenon, mapping its current configurations in relation to the transformation of urban space. It discusses ethical hacking and Living Labs in the neoliberal 'smart city', citizen science and crowdsourcing from the communitarian perspective, and the sometimes explicit policy goal of creating smart districts and 'cool' places with a rising tag on private dwellings. It is argued that the Living Labs strategy fits very well with the deployment of the 'smart city' ideal, responding to two crucial issues the 'smart city' has foregrounded: citizens' more direct involvement and technological solutionism. As a consequence, this strategy has become increasingly popular as a supranational public funding strategy while more often being the outcome of academic and industry joint ventures.

In Chapter 6 I discuss the roles of cities *on the move*, providing a first outlook on policies, processes, and practices which are challenging current mainstream ideas of urban growth and entrepreneurial urbanism. However, the focus on which this 'alternative' vision is based shifts from city to city, with some aiming at complying with current (especially European) innovative legislation on digital rights and privacy and others attempting to reconceptualise the 'smart city' concept altogether in order to present a more coherent alternative to the neoliberal one. I have given different denominations to such cities, which are not conclusive or exhaustive but nevertheless try to encapsulate the variety of visions and frameworks. So, New York City has focused on inclusivity (at least concerning some pilots for Internet access), while Medellin has been experimenting with social urbanism. Amsterdam seems to have a more pronounced interest in preserving digital rights through open access and data, while Barcelona has attempted to change the very foundation of techno-capitalism and reorganise the ownership and management of city technologies.

Chapter 7 debates the possibilities and the limits of a public Internet infrastructure which is municipalised and locally controlled. It focuses on the way in which the Internet and related devices are delivered: whether through public intervention, communitarian DIY, or the market, equal access to unlimited and unmonitored Internet can be thought of in terms of human rights claims, steering policy towards increased fairness and justice in the city of the future. It presents two case studies from the United States, showing a variety of municipalist intervention in the provision of such a critical data and communication infrastructure: the often-time reported phenomenon of Chattanooga, Tennessee, which managed to run a state-of-the-art public service and thus changing its image as bolstering the digital economy, and the more sober pilot run by New York City in the largest public housing complex in the country where a free, although modest, broadband service has been implemented.

Chapter 8 is based on my ethnographic study of an inner-city community of practice born around a Wi-Fi mesh network for broadband provision in the neighbourhoods of Deptford, London. It addresses notions of community and conviviality in relation to the stewardship needed for the technologies to work. I suggest that a 'smart approach' to urban commons and commoning in the 'smart city' can help address some of the above critiques: risk-taking for new technologies and issues of power and inequalities which, ultimately, make for a different meaning of 'smartness'. A 'smart approach' to the commons, this shared space of living and the cultural infrastructure that ultimately make the urban, has to include also the public, most notably the city as a relational scale of urban inhabitation: the innovation and deployment of city infrastructure, the massive roll-out of city-wide solutions to planning and congestion, the coming together of rights to inhabitation and the provision of critical infrastructures can ideally be framed through the pulling of public resources, political will, and citizens' power.

In the concluding Chapter 9, I highlight the politics and ethics of an *alternative* city, suggesting moving beyond the dominant framing reproduced by the 'smart city' advocacy coalitions. Thus, the chapter seeks to advance an ideal of urban future which nests on the theoretic underpinnings of the 'right to the smart city' (Cardullo et al., 2019). This is a bundle of rights that foregrounds citizens as the main actors, who are able to control and shape their spaces of inhabitation. It argues that citizenship ought to be reconfigured along civil and social rights and entitlements, active and 'meaningful' participation, and redistribution of socially accrued benefits for the common good, putting forward notions of citizenship and the scope of 'smart city' initiatives in ways that are thoroughly political. This notion takes seriously 'the urban' as the sociocultural dimension and geographical scale in which people live and work, thus opening to a research agenda which joins the dots of contemporary urbanism with the materiality of dwelling. The notion of 'smartness' that emerges from this discussion concerns, in fact, the complexities and wickedness of city living, fostering an idea of 'the urban' which is very different from the knowable, programmable, and thus linear processes postulated by urban science. Thus, I would rather speak of an 'intelligent city', which is alternative to the current neoliberal one, based on collective rights and entitlement that are deeply political and ethical and an ideal of democratic governance that is grounded on the deliberative power of citizens.

Note

1 https://blog.dellemc.com/en-us/distributed-analytics-meets-distributed-data-with-a-world-wide-herd/

Part 1

The neoliberal city

Reloaded

The scope of the first part of the book is to show how the 'smart city', as currently conceived, purports a set of propositions that are instrumental to the neoliberalisation process of cities and city life. It is argued that citizens in the 'smart city' are thought of as neoliberal subjects that new governance arrangements seek to control and discipline at a distance, while offering top-down technocratic solutions to urban issues affecting *them*. As explored more in detail in the next chapters, the 'smart city' represents the coming together of two rationalities, the science of software and computer studies, and the benchmark-controlled science of urban growth (neoliberal economics). This part of the book looks at the theoretical and epistemological underpinnings of these rationalities, which seek to govern cities and citizens via the power of algorithms. First, I unpack the tight relationship between urban informatics and urban science. Thereafter, I show how this technocratic rationality fits well within a broader neoliberalisation process of urban living. The explosion of big data and their almost ubiquitousness have been the connective tissues through which such a union sticks together. The 'smart city' complements the neoliberalisation process of urban living, offering a mix of technical and neoliberal inspired policy solutions. As such, the 'smart city' has been functioning within an already well-oiled mechanism of city competitions, privatisation of public functions and services, and marketisation of city life. With its baggage of science fiction scenarios and geek culture, the 'smart city' is 'The Neoliberal City: Reloaded' (Wachowski & Wachowski, 1999–2003).

The first two chapters will define the field, providing working definitions for 'smart city' and neoliberal urbanism. This is done by looking at the evolutions of these concepts both from an historical perspective and as contemporary working ideas on which to build new theoretical articulations. Chapter 3 suggests the close fit between urbanism, as an academic discipline and a dimension of governance, and technological development, as the means through which a certain urban future is made available. Chapter 4 presents the current landscape of smart citizenship and governance with case studies from Dublin, Ireland, while Chapter 5 will sum up the new policy landscape made of neoliberal urbanism and smart urbanism, delving into the often-times reported solution to citizen participation: Living Labs.

This set of research offers a panoramic on the 'smart city' as commonly conceived and imagined in public discourse and policy. In this novel space, city administrators' 'smartmentality' (Vanolo, 2014) and the profit-seeking endeavours of high-tech companies merge by relegating the 'citizen' in the intricate, and often invisible, deployment of 'smart city' technologies. This operation pattern works in the background of everyday life while being played more blatantly in the foreground of public speaking and funding seeking initiatives. In this context, it is argued that claims concerning the production of 'citizen-centric' smart cities are largely tokenistic, with city administrations and corporations still owning and controlling urban governance and services, and 'smart city' initiatives being instrumental to technologically led entrepreneurial urbanism (Hollands, 2008; Kitchin, 2015; Swyngedouw, 2016).

1   Smart cities

The 'smart city' has become such a loaded concept in policy discourses, academic papers, and industry reports, that it is difficult to provide an agreed upon definition. The concept has gained great importance in recent years, becoming a buzzword for various interest groups from neoliberal urbanists, to computer science and information systems experts, to industry managers and city leaders. Some scholars highlight technological innovations (interconnected sensors, devices, servers, and algorithms for automated responses and data collection) and the new forms of government these innovations produce by data streams which feed dynamically into management systems and control rooms (Luque-Ayala & Marvin, 2016; Sadowski & Pasquale, 2015; Vanolo, 2014). Others consider primarily the opportunities such technologies offer as spatialised or collective intelligence, in terms of mobility and communication (Foth, 2016; Picon, 2015). Others focus on the socioeconomic environment that favours technological 'fixes' to urban problems and creates start-up hubs and new jobs in the 'creative' sectors of knowledge production (Caragliu, Del Bo, & Nijkamp, 2011). Other scholars emphasise the epistemological underpinnings that relate urbanism to computer- and software-led sciences (Foth, 2017; Kitchin, 2014a). Other critical scholars instead would favour a citizen-centric model that fosters social innovation, civic engagement, and transparent governance (de Waal, 2014; Morozov & Bria, 2018; Townsend, 2013). Clearly, one-size-fits-all is not a feasible path in the 'smart city' discourse.

The scope of this chapter is partly to unveil the 'smart city' as a cultural construction. Here, I want to consider the processes of abstraction that take place when moving, for example, from a handshake occurring between a piece of code and another algorithm (in order to connect, for instance, a smart domestic device to a remote server) to the ideological construction of the city as a 'system of systems', that is, an integrated and connected interface through which data and their algorithmic processing determine the city's optimal functioning (Batty, 2017). Although technological development is also very important in determining what is 'smart' about smart cities, it is my contention that the 'smart city', as a concept, maintains a cultural dimension that goes beyond its technological affordance. The 'smart city' holds on to an ideal urban future which has rapidly become significant in policy and industry parlance (Kitchin, 2019b), something more than its cables, sensors, and servers, or a combination of these. This ideal, according to Vanolo

(2014, p. 889), stems from and feeds into a 'smartmentality', where "cities are made responsible for the achievement of smartness – that is, adherence to the specific model of a technologically advanced, green and economically attractive city." Moreover, I would contend that the shift from the techno-centric city to the 'citizen-centric' city highlights the cultural components that are mobilised, rather than only the technological layers. In fact, there is always an imaginary attached to technological creation, maybe a specific idea of the future or an answer to a particular problem or a generic danger. If we scale this up to the city level, and to the transnational landscape of entrepreneurial urbanism, we can perhaps see the 'smart city' also as a sort of cultural production, a pre-packaged narrative for the current publics.

The curious thing about computer technologies is their rapid rate of obsolescence: I am sure that every reader has experienced the speedy devaluation of their own portable electronics due to continuous upgrading, power enhancement, and size shrinking. Maybe a 2-year old iPod has already been pushed out of production, becoming an obsolete piece of electronic junk for which updates and maintenance are no longer provided (see Gabrys, 2011); or a television display has become so 'smart' that channels disappear without the appropriate subscription to the private provider. These are recurrent experiences of everyday technology. If we scale these up to the entire city, we would probably need to start thinking about the history of computer studies and urban science not just as new horizons of enthusiastic development and affordance, but also as a history of announced obsolescence and decay: this history is reminiscent, for me, of the Arcades in Paris described by Walter Benjamin as 'urban ruins' (1940). It is striking how the history of urbanism in relation to computer-led technological progress maintains that sense of unmatched desires, frustrating attempts, and dream-like status, which elevates beyond technology and infrastructure (sensors, firmware, cables, servers, or data) and becomes a cultural abstraction (e.g., the intelligence of the crowd, the cloud, the apps that 'fix' everything, or the anthropomorphic and caring robot).

I think a good way to disentangle this idea(1), the 'smart city', is thus to reverse engineer the process of its creation; that is, to deconstruct it in order to reveal its designs and architecture, and to extract knowledge from it. Reverse engineering the 'smart city' will allow us to improve our understanding of the underlying source codes and layers that make the 'smart city' of today, and maybe even to expose vulnerabilities and find alternatives, moving from the least abstract to the most abstract layer. Thus, I will take an historical perspective on the 'smart city' which will show the complexity but also the underpinnings, theoretical and epistemological, that sustain the concept. By paraphrasing Tung-Hui Hu's (2015) brilliant exposition of 'the cloud' – the Internet – we can look at the prehistory of the 'smart city' following two directions. First, the 'smart city' becomes one of the winning discursive metaphors for the way contemporary society organises and understands itself, "a cultural fantasy, always more than its present-day technological manifestation" (p. xxiv). The 'smart city' discourse, in fact, has been gaining traction by hinging its rhetoric always on something *beyond* the sole technology: for instance, the future of the planet, or the collective intelligence of the crowd, or the global interconnectivity of communication anytime and everywhere. This cultural dimension has often allowed proposers and deployers of smart technologies to make abstraction from the infrastructural, material, and social aspects which, ultimately, these technologies foster. For Hu, that "everything is connected" – as well as its dystopian twist of 'everyware' (Greenfield, 2013), the absolute surveillance stage – is a product of a system of belief.

The second direction of research is the acknowledgement that developments in computational and analytical software, aided by the vast proliferation of data sets of almost everything, are consolidating a specific way in which social scientists look at the world and, particularly, at city processes. Urban informatics, urban science, and algorithmic rationality represent the fundamental epistemological changes brought in by the proliferation of big data. Moreover, by relying on data-sets and forecasts, they sustain the economic rationality of efficiency, performance, and quantitative deliverables proper of neoliberal economics. As Hu suggests with regards to 'the cloud' (2015, p. xxix), "the technology has produced the means of its own interpretation, the lens through which power is read, the crude map by which we understand the world." The growing affair between computer-aided technologies and neoliberal urbanism will eventually provide us with a working definition of the 'smart city' as the assemblage of various socio-technological initiatives which seek to predict and control city living by way of remote, interconnected, and market-led forms of governance. It will also show how this concept has gained importance in recent years, becoming a buzzword for various interest groups, from neoliberal urbanism advocates to computer science and information systems experts to industry leaders and city managers.

At the end, we ask how the combination of these various rationalities and cultural constructions can be applied to a citizen-centric 'smart city'. And further, in the following chapters: What role(s) is the citizen prescribed to enact in the 'smart city'?

Pre-history

The prehistory of the 'smart city' starts with the early attempts to integrate digital computer technologies in the life of cities, and thus managing (that is, controlling, anticipating, and steering) the latter with the aid of the former. In the early 1950s, computers were first developed and used where solutions were needed to control equipment over long distances. SCADA (Supervisory Control and Data Acquisition) systems were introduced to monitor and control utility infrastructures – oil and gas pipelines, water distribution systems, electrical power grids, and railway transportation systems – and relied on personnel to manually operate buttons and analogue dials (Kitchin, Cardullo, & Di Feliciantonio, 2019). Cities began to centralise populations and functions, but the emphasis was more about scaling up infrastructure networks (usually, to regional or national levels). Utility users had a rather passive role, generally framed within the civic paternalism and hierarchical forms of governance proper of the Keynesian state, although also becoming the beneficiaries of massive public investments in infrastructural projects of large scale, typically owned by the state or its subsidiaries. Of course, computer systems were not portable and relied on machine-specific proprietary software to function: thus, early SCADA systems were independent mainframe systems with no connectivity to other computers (other than a backup mainframe). Counter-intuitively for today's standards of off-grid security and renewed calls for partly siloed technologies (Kitchin & Dodge, 2019), SCADA technology was designed as a centralised control system: this has long influenced the collective imagination with possible grid failure, able to shut down an entire city or region functions (catastrophic blackouts, avoided derailments, hackers' or secret services' attacks from the Cold War scenario).

As a consequence, the goals of securitisation and control of 'the system' determined the future developments of informational models as more decentralised and distributed systems, thought to be more resilient to external attacks (during the Cold War represented by the atomic threat). The ARPAnet (as well as TOR, The Onion Router) have been notoriously developed by the US Defence research, and soon became national priorities. They eventually trickled down to civic implementations: for instance, the Internet. This assumed resilience of the distributed network, a 'network of networks' in fact, soon became the epitome of freedom from top-down cultural models, top-down systems of control, and top-down surveillance. For Hu (2015, pp. 6–7), "this model of rupture remains a seductive myth because it explains the dispersion of power through the formal qualities of the computer networks that supposedly enable it." This set of ideas attached to decentralised systems has lasted till today, with some activists misunderstanding capitalism as necessarily attached to a centralised system rather than a model though which socioeconomic relations operate.

As Hu (2015) further suggests, "this seemingly distributed network is built on top of a layer that can only be *centripetal* in nature" (p. 14, original emphasis). His analysis looks closely at the economic geography of the Internet provision in the United States, from the data centres and the cabling systems. The privatisation of Internet provision and the fact that new infrastructures are layered on top of old ones, so that "virtually all traffic on the US Internet runs across the same routes established in the nineteenth century" (Hu, 2015, p. 7), determine an oligopoly in most countries (Dodge & Kitchin, 2003). In addition to this material layer, forces of centralisation have been working on the cultural level too. Recent cultural constructions of safety and security have both cemented a border narrative of the network (e.g., against attacks from Russian hackers or Nigerian spammers), while at the same time endorsing the self-reliant behaviour of the responsible user (e.g., *vis-a-vis* pornography and piracy). This has led Hu to conclude that, with regards to user participation, "the cloud is a neoliberal fantasy" (2015, p. 145).[1] With the smart technologies, we face the same challenge today – and partly this book will seek answers to this: are integrated and distributed communication systems, IoT devices, and city platforms enabling citizen participation and power?

By the 1970s, cybernetic thinking led some to recast the city as a 'system of systems' which could be digitally mediated and optimized (see Kitchin, Cardullo et al., 2019). Jennifer Light (2002, p. 608) has observed that "techniques and technologies from military operations research, such as systems analysis and computer simulations, started offering a new direction in city administration," with planning and management quickly emerging as investment targets for the military-industrial complex. For instance, the RAND Corporation, an Air Force think-tank of the US Department of Housing and Urban Development (HUD), used management models in New York City to calculate and optimise the travel time of fire service. As a consequence, by the mid-70s fire stations were forced to close or relocate from high-density neighbourhoods such as the Bronx (Wallace & Wallace, 1980). The point of relevance here is that the RAND Corporation's analytic models were validated by a computational rationality, "only by comparison with a simulation model" (1980, p. 419). This placed statistically calculated efficiency before empirical observations or policy concerns, at a time when no other service was putting budget cuts in place.

Parallel to increased interconnectivity, the development of microprocessors in the 80s and 90s allowed computers to become much smaller and more personal machines, while the software industry made their operating systems and files portable. Mobile phones made their entry, initially as status symbol for 'yuppies'. It was a time for great optimism, even euphoria, around the digital since technological change, the growth of markets, and increasing competition were leading to rapid cost reductions and greater performance (Graham & Marvin, 1996, p. 28). Technological development of the Internet and the World Wide Web seemed unstoppable and destined to grow massively. At a more subtle level, microprocessors became embedded components of devices and systems ranging from cars to digital cameras. This was bringing a crucial change in the everyday life of cities – maybe even a more important one than the digital communication revolution – to the point that "digital intelligence was no longer tightly concentrated, but was now ubiquitously present throughout urban environments" (Mitchell, 2007, p. 4). Scholars produced an array of literature debating the rising of, and the hype around, the 'digital city' – although the term is originally attributed to Ishida and Isbister (2000), here it is used as a generic denomination trying to capture the essence of this chapter, that is, the progressive and dialectic intertwining of digital and computational processes and technologies with the urban.

This interrelation was perceived at the time as a dramatic change in the communication and information networks, seen as an opportunity for freedom through the dissolution of cities via decentralised networks of small-scale communities (e.g., Castells, 1996); or, as a new social and spatial dimension (Mitchell, 2007). For others, the digital city offered an additional splintering and exclusivity of service provision (Graham & Marvin, 2001); and for others still, it became a source of continuous technological innovation, thanks to the "feedback loop of digital sensing and processing" (Ratti, 2009).

The narrative of a 'virtual' world (a cyberspace) increasingly separated from, or opposite to, the material one seemed to dictate both the cultural imagination of freedom fighters (critical scholars, cyberpunks, activists, bloggers) and that of industry and city managers. For Castells (1996, emphasis added, p. 442), physical places are now held in a dialectical relationship with flows, these being "the purposeful, repetitive, *programmable* sequences of exchange and interaction" between physically disjointed positions held by social actors. The possibilities of personal computers connected by a spreading network became also extremely vast for cities, in both the real and imaginary realms. For instance, 'virtual cities' became early attempts to create digital copies of cities and city functions in cyber-space, either as strategy games or as GIS and basic 3D visualisations (Graham & Aurigi, 1997; Willis & Aurigi, 2017). Examples of these virtual cities are GeoCi-ties,[2] SimCity[3] and, later on, Second Life.[4]

At the same time, large investments in e-government followed with the delivery of services and interfacing with the public via digital channels and e-governance techniques for managing citizen activity via digital tools. Willis and Aurigi (2017) suggest the 'digital city' became mostly linked to the use of ICT in public administrations and with the e-government practices, in order to create easy access to information about the city and its services and to experi-ment with participatory practices. Despite the relative availability of comput-ers connected to the Internet, however, substantial delivery of urban services seemed to have lagged behind. In other words, the promises of the Internet con-necting a community of strangers to the city remained a vague and impracticable objective for many. On the other hand, connectivity extended the networking of infrastructure, such as the broad adoption of traffic management systems and surveillance cameras (Lyon, 1994).

Towards the 'smart city'

Parallel to the new ways in which people became able to communicate with each other – and getting empowered by way of expressing their own views over the Internet (blogs and personal websites dominated in this period, before social media platforms were developed) – a more subtle change was happening through a new wave of technological innovation: "Minuscule digital cameras and microphones gave the Internet eyes and ears everywhere. GPS and other location technolo-gies made devices, such as automobiles and mobile phones, continuously aware of where they were. RFID tags embedded in products and packaging began to revolutionize logistics and retailing" (Mitchell, 2007, p. 4). The Internet, initially built for people, suddenly became the vector for the development of millions of other devices, objects, and machines. For Mitchell (ibid.), this integration of func-tions and data made cities "intelligent with their own artificial nervous system," such as "digital telecommunication networks (the nerves), ubiquitously embed-ded intelligence (the brains), sensors and tags (the sensory organs), and software (the knowledge and cognitive competence)."

The catching metaphor of the city-body is not new, of course, and still has much traction today. This idea fills imaginary city flows – of traffic, goods, and information – that was already familiar to Haussmann's intervention in Paris and Moses's in New York (Sennett, 1992), while for Lewis Mumford (2009, p. 363) highway building was "a tomb of concrete road and sand ramps covering the dead corpse of a city." Victorian planners had already thought of the underground train system as the limbs (peripheries) converging to the heart of the city-body (the centre of London) in order to take care of the 'disease': a "great thrombosis of traffic which clogged the highways that were the veins and arteries carrying the city blood" (Douglas, 1963, p. 13).[5] Similarly, visions of organic city-body feed urban regeneration policies imagined as curative interventions to terminally ill city's remote quarters and public estates (Keith & Pile, 1993). These anthropo-morphic visions of city space and its connecting infrastructures resurface today with regard to the 'smart city': this is often considered a superb container of human interaction and communications. For instance, Townsend maintains that the digital city is a "living organism," "alive with movement." One of the questions he asks is, "How do the cells of the city cluster to form tissue and organs?", that is, "How do various systems communicate and interact with each other?" (preface to Foth, Klaebe, Adkins, & Hearn, 2009, p. xxviii).

That cities can 'sense' the surrounding environment and 'think' an appropriate response – sometimes even making the necessary adjustment as an automatic act that does not require human intervention – is, however, something new. According to Featherstone (2009, p. 4), devices powered by radio-identification sensors are creating "an animate environment with agential and communicative pow-ers." Once provided with ID tag, computer chips and a transmitter/receiver, these objects are no longer fixed and silent but part of an urban ecology played between the external environment and devices interconnected with their remote databases (Hayles, 2009). For Shepard (2011), this is now a 'sentient city' where informa-tion and live data feeds become part of the material city. Fast forwarding to the 'smart city' of the near future, we can see how the metaphor of city-body has gained more traction in the collective imagination. For instance, high-tech giant Huawei's new city platform, which integrates feeds of data from devices with AI processing, is thought as the city's "nerve system with effective self-learning and self-development" capabilities.[6]

This novel abundance of data is due to the coming together of different deter-minants of digital computing and city functioning. Following Kitchin (2014a), we can group these into: technological factors (sensor- and software-enabled urban environment and storage of data at affordable costs); the explosion of personal devices networked via social media and Web 2.0 applications; and advances in database design and in systems of information management. The increased impor-tance of data in the very fabric of city living is crucial to understanding the pas-sage to our latest sociotechnical and cultural creation: the 'smart city'. While data collection and analysis have long been tools used by urbanists, data sets such as the census of populations were "a relatively limited sample of data that are tightly focused, time and space specific, restricted in scope and scale, and rela-tively expensive to generate and analyse" (Kitchin, 2014a, p. 3). Big data are the exact opposite: they are incredibly huge in *volume*, high in *velocity*, and diverse in *variety* (the three 'Vs' of big data). Additionally, Kitchin notes that big data are also exhaustive in scope (samples can be approximate to population size), fine-grained in resolution (data can be indexical and detailed), relational and flexible (connected to other datasets, but also scalable). Thus, big data are dynamic and interconnected datasets that address the city as a quantifiable and knowable entity, which is thought to be manageable in real-time.

A detailed discussion on big data is beyond the scope of this book, although the salience of big data for smart urbanism needs to be highlighted here. This is, in fact, a crucial step for the making of smart cities and for the roles circumscribed to their inhabitants, the 'smart citizens'. While the 'sentient' or 'intelligent' city was still based on a model of urbanism *informed* by data feeds, the 'smart' city fosters a model of urbanism that is data-driven at its roots (Kitchin, 2016b); it is *produced* by data feeds. This means that the 'smart city' development has become instrumental to the implementation of devices and infrastructures which, ultimately, produce big data. As I noted in the Introduction, virtually every dimension of life in cities (and the life of cities) is now emerging as fields of digital interaction and data production, circulation, and analysis. Thus, for Kitchin (2014a), big data have become the 'smart city' "mode of production", used in order to know, predict, inform, and deliver its policy and services.

On the one hand, in fact, what makes a city 'smart' is the new modality of real-time urban governance, which is "highly responsive" to big data systems (Kitchin, 2016b). Congestion monitoring systems (Kitchin, Coletta, & McArdle, 2017), control rooms (Luque-Ayala & Marvin, 2016), and city dashboards (Kitchin, Maalsen, & McArdle, 2016), are now spanning citywide systems which draw together data streams into a single data analytics centre. The panoptic view that big data-driven urbanism wants to achieve is evident here. However, Coletta and Kitchin (2016, p. 3) suggest two different types of algorithmic governance in the 'smart city': "the first is more interventionist, adaptive and direct in nature, wherein the on-going management of a city system is over-determined by code [e.g., traffic management systems]; and the second is more contextual or performance management orientated, wherein the rhythms are measured, monitored, recorded and modelled, with interventions more periodic and indirect [e.g., city dashboards]."

On the other hand, big data systems are "setting the urban agenda and are influencing and controlling how city systems respond and perform" (Kitchin, 2016b, p. 3). Another determinant for the current dominance of big data – in addition to the already mentioned technological factors, explosion of personal devices, and advances in database design – is, in fact, the rise of new forms of analytics designed to cope with such an abundance of data. In order to show the stringent relevance of big data to smart urbanism, we need to take into account that "big data analytics requires expertise from mathematicians, statisticians, computer scientists working on the algorithmic challenges of big data" (Foth, 2017, p. 5). This expertise sits on the methodological and epistemological layers drawn within the interdisciplinary field of data science, and more specifically of urban informatics/science. These layers are concerned with big data, and move towards a new form of empiricism and a new mode of data-driven science which is epistemologically inductive (Kitchin, 2014a). The next section looks at the increasing connections between algorithmic governance and the field of urban informatics/science.

Algorithmic governance and urban science

System thinking is at the roots of urban science, an interdisciplinary approach that uses statistical analysis and data analytics, including spatial statistics, remote sensing, data mining, simulation-based modelling, volunteered geographic information, cell phone data, machine learning, visual analytics, and simulation (Helbich, Jokar Arsanjani, & Leitner, 2015). The scope of system thinking is to "determine urban 'laws', conduct real-time analysis of systems, produce new theoretical insights, develop a synoptic and integrative science of cities, and to translate the knowledge produced into practical application" (Kitchin, 2017, p. 2). The 'computable city' provides the ideal environment for the conception and application of the disciplinary streams of urban science. With cities becoming complex and infrastructural systems made of a 'constellation of computers', it appears much more appropriate and easier than ever before to address, simulate, and predict a variety of urban problems via computational methods.

For instance, big IT companies like IBM propose a 'smart city' model based on a 'system of systems'. This presents a city operation system (civic processes and governance); city user systems; and city infrastructure systems (energy, water, communications, and transport infrastructures) (Söderström, Paasche, & Klauser, 2014). Chinese technology giant Huawei has recently launched a 'smart city' platform that integrates AI, Internet of Things data and cloud technologies, thus providing a "vast operating system for cities".[7] Similarly, Microsoft has announced its own platform, Azure Digital Twins, for coordinating under one system an announced proliferation of simulated environments and data from Internet-connected devices (IoT).[8] Initially developed by NASA for space simulation, the virtual reality of the 'Digital Twin' is a digital copy of any actual or planned product, which is fed in with live data from both the real object (its hardness, for instance) and the algorithmic learning of the virtual model itself (e.g., simulated shocks). The 'Digital Twin' is therefore a dynamic and iterative concept: it is a virtual copy running in the cloud that gets richer with every second of operational data. Some big players in the field of software simulation for Industry 4.0 are now thinking to scale up the experimental products and processes of the 'Digital Twin' to the city level. One of the world's leading suppliers of intelligent automation solutions, Kuka, is working with the city of Singapore to construct its 'Digital Twin', a "multiple, dynamic, and interconnected 3D experience of city space, roads, infrastructure, traffic, etc., which can simulate variables such as movement of people and wind direction" (representative's intervention at Industry 4.0 conference, Cork, Ireland).

This is not a return to, or an improvement in, the design of the 'virtual city'. That was, in fact, a fictional construction of city life: a game in which players were able to build hospitals or increase police presence, or buy new clothes for their avatars. It was *another place*, a parallel cultural but a distinct dimension from the real city. What city modelling and simulation are offering today is rather a new form of 'abstract urbanism', based on the knowledge that models are imaginary "without being either merely false or simple reification" (Fuller, 2017, p. 47).

Its tight connection with the real object is the paradox through which this new 'virtual city' works. This is because the data feeding from sensors implemented on the real object are the ontological foundations in the life cycle of the 'Digital Twin': from design, programming, commissioning, and operation to simulation, each phase of the process is informed by and produces data. The availability of diverse and massive data sets, whose *modus operandi* is "purely inductive in nature" (Kitchin, 2014a), has been decisive to the enforcing of this calculative rationality, modelling, and simulation of city functions.

On the one hand, this epistemology is fostering a new form of empiricism, where data are generated and analysed generally free from theory: data are considered a reflection of the world 'out there'. They speak for themselves, free of human bias, and anyone with analytical skills can extract laws and trends on how cities work. On the other hand, this inductive epistemology is the strategy adopted within data-driven science, which "uses guided knowledge discovery techniques to mine the data to identify potential hypotheses, before a traditional deductive approach is employed to test their validity" (Kitchin, 2016b, p. 4). A problem with inductive methods is, of course, the fact that (big) data ultimately determine the research questions, which are often adjusted to the available data set, or to seek an answer to a question that it was never designed to answer in the first place (Kitchin, 2014a, p. 9).

Thus, as Kitchin (2016b) notes, there is a "strong recursive relationship" between data-driven urbanism and urban science/informatics. While data and fields of application are taken for granted within the inductive epistemology of data-driven urbanism, urban informatics and urban science develop conceptual frameworks (e.g., informational and human – computer interaction) and tools for analysis (e.g., computational modelling and simulation) in order to analyse such data for city-wide solutions. Therefore, urban science and urban informatics seek to address "the two fundamental challenges posed by urban big data": how to make sense of them and, at the same time, how to generate new urban 'laws' and actionable outcomes (Kitchin, 2016b). While Foth (2017) maintains that the primary scope of urban informatics is to harvest the opportunities offered by ubiquitous computing (he includes all aspects of the "convergence of physical and digital aspects of the city," thus also urban media), for Kitchin (2016b), methods and epistemology are crucial in defining data-driven urbanism, that is, a computational understanding of city systems that reduces urban life to "logic and calculative rules and procedures."

As a consequence of the coming together of algorithmic governance and the calculative rationality of urban science, the roles citizens take in the 'smart city' are thought to remain circumscribed to urban environments that are predicted to behave and adapt within this computational rationality (Gabrys, 2014): citizens are data-points, simulated variables, and users-customers. Human behaviour in the urban environment is simplified, literally codified and rendered knowable through analyses of systems and services which they cannot grasp – it is the scope of algorithmic processes and the epistemic community around them to do so – let alone influence.

Concluding remarks

Algorithm-led processes and technologies at the heart of the 'smart city' matter enormously. This is because their working embodies a process of 'translation' of how cities function; at the same time, the actions they generate, sometimes as anticipatory forms of governance or adaptive behaviours, reshape city life. The latter process is called 'transduction' which recursively integrates the former: "software and the work it does are the products of people and things in time and space, and it has consequences for people and things in time and space" (Kitchin & Dodge, 2011, p. 13). In the examples of the 'Digital Twin', we see the deployment of a dynamic and iterative process, the 3D model of reality (translation) which, in turn, can inform changes in the original product (transduction). By bringing under their own 'cloud' this variety of data, high-tech giants aim to collapse the city and its functions into an interface which is both ubiquitous and continuous: here, executive elements of the platform are consequential to its knowledge acquisition system, but the system can learn from and adapt to its own analyses and responses. Thus, the solutionism of urban science mixes with the problem-solving logic proper of computer science: computation is being utilised to *both* understand the functioning of cities *and* manage and control the urban systems that provide the valuable data and the predicted responses (Fuller, 2017).

When this dialectic is extended to the city-wide scale, urgent questions around governance of citizens and processes emerge: Who has the ability to develop and manage such complex data-driven systems? Who owns or has access to valuable and strategic data? And, since urban informatics systems are increasingly 'smart', that is designed to be automatically and autonomously responsive or adaptive, who is accountable for their unwanted effects? In other words, if software, AI, and smart technologies have some sort of agency, in that they have an essentially executive state, what kind of ethical and political oversight can be implemented? These questions are not exclusive, but all are very relevant to this new way of computing cities. My argument is that this debate belongs to the ethical, legislative, and philosophical domains, not only to the technical one; ultimately, it is an inherently *political* debate. Unfortunately, the present configuration of 'smart cities' bolsters a model of development which is led by market principles of efficiency and patents, close data and black-boxed algorithms, private industry providers and competitive edges, simulation modelling and predictive governance, technocratic advisors and experts. In other words, the prevalent logic of the current 'smart city' is a post-political ideal that fosters civic paternalism, in regard to the needs of cities, and stewardship, in relation to implementation and uses of smart technologies.

The scope of this book is to slowly unravel these issues and start answering the above questions by keeping a close eye on the 'smart citizen', between cultural constructions and their daily practices of inhabitation. 'Digital city' dwellers have been understood mainly as Internet users, with alternate fortunes. Depending on who owns this cultural imaginary, they have been framed as cyberspace travellers, model citizens, subjects of surveillance, or independent cultural producers. 'Smart citizens' of today owe a lot to these cultural configurations, becoming symbolically dominant in public discourse, but marginal, and mostly passive in practice. What needs to be noted here, in fact, is that the fascinating and fast-developing history of computing technologies in cities had a crucial turning point when users of personal computers (mostly located at home or in their offices) turned into individuals always connected to networks or 'clouds' via their mobile devices (Fuller, 2017). The novelty of smart devices is that these appear as personal and portable but, at the same time, are also *always* interconnected with platforms, systems, and cloud spaces which are very centralised. Thus, on the one hand, mobility and consumerism can become ad-hoc, always on-time, and correlated variables of 'smart citizens' in time and space. On the other hand, the move from the personal hard drive to the 'cloud' means that the "user directly faces platform-based services that are constantly optimised to compete with other services and to capture potentially valuable data on the user's interactions as part of a population of representative and emergent types" (Fuller, 2017, p. 214). The current data extraction practices represent an inversion of the use-exchange value dialectics: while for Marx, the exchange value of the commodity hides commodity's utility (its use value), secretive platform algorithms hide the price (the exchange value) of users' voluntary contributions (likes, shares, etc.) (Skeggs & Yuill, 2018).

In other words, there is a continuous interplay between opacity and transparency, which seems to lean currently towards 'smart citizens' roles being mostly performed as consumer, resident and, above all, data-point. According to Kitchin (2014a), the variety of modes and practices for the making of 'smart city' initiatives has generated confusion in terms of the relative agency and control citizens have in their smart environment. This confusion seems to persist in many narratives of the 'smart city', focusing on everyday direct experience of users' data on social media rather than more subtle and relentless practices of data extraction. While the former is mostly volunteered, in fact, the latter are direct (e.g., surveillance) or automated (e.g., inherent to the functioning of the system) acquisitions over which the 'citizen' has very little or no control. Chapters 4 and 5 will present a round of empirical research locating 'the citizen' around Smart Dublin initiatives and those projects labelled as 'citizen-focus' under the European Commission programme for smart cities and communities (EIP-SCC).

In concluding this chapter, I would like to point out another key difference between what we have seen as history of digital computing deployed in cities and the current 'smart city' movement. 'Smart cities' are profoundly urban, in the sense that their data-driven mode of production relies on the density of occurrences, connection points, and numbers of clicks, as well as massive movements of vehicles and people: all things that are very urban and due to the critical mass of users and consumers that dwell in cities. Ultimately, the 'smart city' ethos is very much about making cities attractive places to work and live, because they are manageable, profitable, and they are growing. This is indeed what cyber, virtual, and digital cities *were not*. According to Graham and Marvin (1996), those earlier conceptualisations were infused by a "strong anti-urban emphasis." Advances in communication and on-line virtual worlds were thought to liberate people from the density and externalities of cities, orienting them towards small communities of interests. The 'smart city', in contrast, wants to capitalise on every aspect of daily life, here and now; it has become the social factory 4.0, the embodiment of neoliberal techno-capitalism. Techno-capitalism too is interested in urban life, where it can experiment with biopolitics and real-time governance, an advanced stage in the neoliberalisation process of cities (e.g., Rossi, 2017). Ultimately, it is in cities where new markets for technology can be created (Kitchin, 2015). The next chapter will unpack the coming together of the 'smart city' algorithmic rationality with the tenets of efficiency and market priorities proper of neoliberal urbanism.

Notes

1 I will explore user's roles further, in Chapters 3–5, in the context of a discussion of citizens' participation and power in the smart city.
2 Since 1995, new members created their own webpage choosing to which neighborhood they wanted to belong. Acquired by Yahoo! at the pick of the dotcom bubble in 1999, it was controversially dismissed in 2009, after 15 years and 38 million user-built pages (adapted from Wikipedia).
3 The original game was developed in 1985 with many subsequent versions: interestingly, the player develops a city from a patch of undeveloped land and controls where to place infrastructures, landmarks, and public services, as well as determines the tax rate, the budget, and social policy (adapted from Wikipedia).
4 Project started in 2003. By 2013, *Second Life* had approximately one million regular users, also called *residents*, who could create virtual representations of themselves, called *avatars*, and are then able to interact with places, objects, and other avatars (adapted from Wikipedia).
5 Ironically, London Underground iconic map has been recently redrawn in a way that shows declining life expectancy in the class-divided capital (Cheshire, 2012, p. 12); a staggering 9 year figure is lost between Canary Wharf and Deptford, from one river bank to the other.
6 www.verdict.co.uk/huawei-smart-city-platform/
7 See, www.verdict.co.uk/huawei-smart-city-platform/
8 https://blogs.microsoft.com/iot/2018/09/24/announcing-azure-digital-twins-create-digital-replicas-of-spaces-and-infrastructure-using-cloud-ai-and-iot/

2   The neoliberal 'smart city'

In the previous chapter, I noted the "strong recursive relationship" (Kitchin, 2016b) between data-driven urbanism and urban science/informatics and how this is becoming the established foundation for urban policy – to the point that parts of cities are being reconfigured as 'Digital Twin' models. While data-driven urbanism takes for granted data as the manifestation of urban flows and needs, urban informatics develops conceptual frameworks and tools for analysis in order to make sense of such data for city-wide solutions. In this chapter, I take a step further by showing the interconnections between the data-driven 'smart city' and the tenets of neoliberal urbanism, the organisation and management of city life based on the marketisation of service provision and social relations. The chapter is organised around a discussion and a working definition of neoliberal urbanism, collating its common discursive and ideological underpinnings with regards to the 'smart city' movement, while also highlighting its variety with regards to policy translation and implementation through different scales and places.

One useful entry point for understanding the interconnections between urban informatics, data-driven urbanism, and neoliberal urbanism is provided by Matt Fuller's (2017) reading of Friedrich Hayek's theory of how the market works (e.g., Hayek, 2011). This is classic neoliberal thinking, which centres on the mechanism that allocates resources, wages, and profits. Ultimately, Hayek suggests that this mechanism manifests itself in the price system that regulates individual transactions: a rational agent, so the argument goes, will always make predictable choices to the extent that he or she holds the necessary information which permits such reasoning to take place. It follows that agents who possibly know everything are able not just to accomplish optimal and rational decisions. They are also able to make competitive gains on less informed agents, knowing what and how it happens in real-time (prerogative of big data analytics). As Fuller remarks, "money, understood as pricing signals, finally means we can rightfully subsume economics under media theory" (p. 220). There is an intrinsic truth to this: information systems, stochastic variables from all-encompassing models, and AI experimentation with massive datasets, imply exactly this shift of perspective: a connection between epistemology (how we know things) and future gains (how we put this knowledge in place in a way that is profitable only to us). This is because we know more (in a way of having more data) and therefore we are able to make better predictions (in a way of processing these data faster or more efficiently than others). O'Neil (2017) calls this potential force of algorithmic reasoning applied to speculative financial gains "Weapons of Math Destructions." She suggests that algorithmic rationality is all but an objective and neutral reflection of the reality out there, with choices programmers make about which data to pay attention to and which to leave out. "Those choices are not just about logistics, profits, and efficiency. They are fundamentally moral" (p. 171). But neoliberal ideology defines people *as if* they are always trying to maximise their own wealth and power at the expense of others and everything else. In the second part of this book, instead, I will open to a new ideal for the 'smart city' yet to come based on adjustments and configurations that are intrinsically ethical and political, a framework which is a necessary *alternative* to the neoliberal ideology of the market.

While platform capitalism is actively seeking maximum transparency from individual agents in order to dig and extract as many data as possible (from their shopping behaviour to their intimate life), the other side of this coin is that digital platforms act with opacity and ambiguity. Since these platforms are oriented for individual gains and private profit, they conceal the algorithmic calculus behind their apps, delivery choice, tariff allocation, credit ratings, and so on. The determinants of market prices are therefore as if "everything is a black box, with prices as inputs and outputs" (Fuller, 2017, p. 220). All that agents (consumers) can see are price fluctuations here and now, from their holiday flat to their next cab journey to town. Prices are thought to signal the underlying forces of the market whose 'invisible hand' is however black-boxed in mathematical abstractions, algorithmic imagination, and stochastic thinking. We can see the app interface but we don't know how the calculation to bring us there is made possible. We are delivered credit scores and potentially life-changing financial decisions but we have no idea how this has been computed (Kitchin, Graham, et al., 2019). Black-boxed algorithms can indeed give a marginal edge against competitors because whoever makes the better guess wins: it follows that platform economies want to bring transparency and disrupt established markets by way of concealing the algorithmic rationality behind it (we know very little of how Uber calculates car travel fees, or the gains that Airbnb makes in an individual city, or the computational logics behind fake news recurrence on Facebook).

Information (increasingly in the form of big data) is all that matters in order to get 'the feel' of the market. As Shoshana Zuboff in her book *The Age of Surveillance Capitalism* (2019) has recently suggested, this is what data-platforms and high-tech providers are ultimately seeking: "They sell prediction products into a new marketplace. There are a lot of businesses that want to know what you're going to do, and they're willing to pay for those predictions."[1] As a consequence of this reasoning, everything ought to be computable, metered, and quantified in order to be better predicted. As Fuller notes (2017, pp. 173–179), knowledge is all in such a formulation; but, crucially, this knowledge is translated in the grammar of the market: "purchase, sell, wait, watch, track and calculate." Hayek imagined the market as a sort of interface which individual agents, consumers, or producers, can access with wonder, adjusting their preferences and thus making rational decisions by following price fluctuations, like "an engineer might watch the hands of a few dials." Atomised individuals strive to pursue and maximise their own self-interest in the name of liberty, a contested liberal right claim that safeguards individual capacity to access resources for their own benefit. Theorists of individual freedom go further by claiming that even democracy can be an obstacle in the way of economic liberty, thus risking becoming an impediment to (market) freedom. In "The Constitution of Liberty" (Hayek, 2011) this is clearly expressed: democracy is simply a "method, indicating nothing about the aims of government" (p. 168). For instance, in terms of its will at maintaining and protecting individual liberty which, for Hayek, represents the supreme goal of society: "The prospects of liberty would have little chance of surviving if we relied on the mere existence of democracy to preserve it" (p. 173). As a matter of fact, it took a military coup by Pinochet in 1979 and years of violent tyranny in order to force the first wave of neoliberal orthodoxy upon the people of Chile. Thus, a neoliberal- oriented policy can shift power away from the democratic institutions and the political decision-making process; and established rights towards external and less accountable political-economic assemblages: for instance, the technical expertise of central banks, off-shore corporate subsidiaries, or the IMF. While the ideological foundations of neoliberalism suit perfectly a rational agent moving in a marketplace made of available information and data, it matters enormously that cities have been very attractive sites to capital. With their density of infrastructures, business, and people, they are "major basing points for the production, circulation, and consumption of commodities," as well as themselves being "intensely commodified" sites (Brenner, Marcuse, & Mayer, 2009, p. 178).

The connections between capital and power in driving the processes of urbanization, and in reproducing sociospatial structures and power relations in cities, have been explored by a vast literature. This has been focusing, for instance, on the circuits of capital accumulation (Harvey, 1978), the operations of neoliberalism (Peck, Theodore, & Brenner, 2009), the heritage of colonialism (Robinson, 2006) and the implications of state governance (Massey, 2007). Thus, neoliberal urbanism can be summarised as a model of urban growth based on marketisation, that is, the further "subordination of place and territory to speculative strategies of profit-making at the expense of use values, social needs and public goods" (Peck, Theodore, & Brenner, 2013, p. 1092). In a neoliberal framework, the market arranges services, infrastructure, and resources (including housing and public space) that have been thus far provided by the state. Such a shift in the ownership of what were public assets (privatisation) and provisioning of services (marketisation) has been driven by arguments concerning efficiency, competitiveness, and value-for-money that paved the way to strong austerity policies (Peck, 2012). At the same time, financial capital, increasingly central to innovation-led growth, has been strengthened through market re-regulation which protects short-term and risk-averse returns (Lazonick & Mazzucato, 2013). Cities have been not just the sites of production and experimentation of technologies and their ultimate target market. They have also been privileged sites for the experimentation of neoliberal policy (Rossi, 2017), thus becoming "increasingly central to the reproduction, reconstitution and mutation of neoliberalism itself since the 1990s" (Peck et al., 2009, p. 50). As a historically grounded project, the 'smart city' appears to be the latest attempt to use and reconfigure the city as an accumulation strategy. For instance, Dublin in Ireland illustrates this phasing, adopting ideas of entrepreneurial planning in the 1990s, the creative city discourse in the 2000s, and finally the smart city in the 2010s (Coletta, Heaphy, & Kitchin, 2018; MacLaran & Kelly, 2014). The 'smart city' forms a particular version of entrepreneurial urbanism grounded on technology design and adoption (Hollands, 2008): through its agenda, private interests seek to capture public assets and services by offering technological solutions to urban problems (e.g., congestion, emergency response, utility and service delivery). While setting appropriate goals for cities via systems of urban benchmarking, the neoliberal 'smart city' aims to attract foreign direct investment, offering areas of the city as testbeds to pilot new technologies, fostering innovative indigenous start-up sectors or digital hubs, and attracting mobile creative elites.

Through the present rolling out of the 'smart city', a double movement has therefore been occurring, or accelerating, at the level of the individual: a process of neoliberalisation of personhood *and* the shifting role from citizens with rights and entitlements to consumers regulated by market laws and contractual small prints. As individuals are moving through a marketplace of 'choices' as consumers, their behaviour becomes a crucial concern for both corporate sellers and public governance: as rational agents, in fact, their behaviour can be both predicted and nudged. On the one hand, this predominantly market-led framework fosters ideals of efficiency and personal and corporate gains rather than the common good, thus making concealment of information a rational market strategy for success. On the other hand, the 'smart city' presents the dashboard scenario of an all-knowing city which depends heavily on the extraction of as many datasets as possible. What smart urbanism is ultimately seeking, then, is the dashboardisation of the city life in an all-encompassing and real-time way, de facto acting forms of corporate and state surveillance on individual choices. This surveillance is necessarily operated through data extraction mechanisms and practices which eventually feed predictive modelling and forms of anticipatory governance.

In sum, critical scholars consider the 'smart city' as the ultimate phase of a neoliberalisation process of cities and urban life, although a stage led by techno-capital and advancement in the design and production of interconnected, cloud-based, app-implemented, and (above all) algorithm-led technologies of data extraction, analysis, and prediction. The present and the following chapters are both an exercise of academic thinking around the vast work critical scholars, especially urban geographers, have produced on this topic and an empirical excavation of the current panorama of 'citizen-centric' smart cities initiatives at the European level, with a focus on Dublin: the former draws on the author's research around the initiatives funded by the European Innovation Partnership for Smart Cities and Communities (henceforth, EIP-SCC) (Cardullo & Kitchin, 2019b), while the latter is centred around the work done by the team on the ERC-funded project

The Programmable City (Maynooth University Social Science Institute) between 2012–2018 (e.g., Cardullo & Kitchin, 2019a; Coletta et al., 2018; Kitchin, Coletta, Evans, et al., 2017; Perng & Kitchin, 2016).[2]

Neoliberal urbanism

There is a general consensus among critical urban scholars around the varieties of forms and applications of the neoliberal discourse and related urban policies (e.g., Karvonen, Cugurullo, & Caprotti, 2018; Peck et al., 2009). This section, however, tries to group some of the main recurrent characteristics of the neoliberalisation process which models contemporary urbanism. In this way, it will show how the ideological premises of neoliberalism in general are corralled within urban policies of urban growth and the mainstream 'smart city'. The reminder of the chapter maps the ways in which neoliberal urbanism and the 'smart city' ideal have been colluding, for instance as the backdrop for testbeds initiatives, funding mechanisms, discursive constructions, and procurement processes that characterise the variety of 'smart cities'.

Talking about neoliberalism in general is therefore not an easy task since critical urban, economic, and development geographers have explored and exposed its inherent contradictions, diversity of implementation, and place-specific characteristics. They paid attention, for instance, "to the different variants of neoliberalism, to the hybrid nature of contemporary policies and programmes and to the multiple and contradictory aspects of neoliberal spaces, techniques, and subjects" (Gibson-Graham, cited in Castree, 2006, p. 2). Much more than a theoretical and a geographical abstraction, neoliberalism has given way to a plurality of "actually existing" processes and practices (Brenner & Theodore, 2002). Thus, critical scholars prefer to talk of *neoliberalisation*. This is a sort of "mid-level theory" which, on the one hand, presents a distinct kind of policy whose environmental and social impacts can be fairly understood (Brenner, Peck, & Theodore, 2010; Castree, 2006); while, on the other hand, it focuses on "a historically specific, fungible, and unstable process of market-driven sociospatial transformation, rather than a fully actualized policy regime, ideological apparatus, or regulatory framework" (Peck et al., 2009, p. 51). According to Peck (2004, p. 395), while the neoliberal discourses and strategies that are mobilized in different settings share "certain family resemblances," local institutional context "clearly (and really) matters in the style, substance, origins and outcomes." I attend to this double call, highlighting "family resemblances" of neoliberal smart urbanism, while referring to case-specific 'smart city' initiatives whenever possible.

On the one hand, there is a risk of reifying neoliberalism as a hegemonic and unifying entity and thus to exaggerate its power. What counts as 'neoliberalism' does not appear to be a matter of consensus among critics either. In many cases, privatisation and marketisation are the key criteria; in other cases, additional features are emphasised such as governance or financialisation. Moreover, neoliberalism comes in a variety of means, shaped by national and local political economies, political ideology, state policies, institutional cultures, market practices, legal frameworks, and public sentiment (Brenner et al., 2010); it does not operate in all places at all times in a unified, universal manner, but has varying stages, topographies, and topologies (O'Callaghan, Kelly, Boyle, & Kitchin, 2015). Therefore, the question around neoliberal urbanism can shift to how policy travels, is adapted, and affects cities differently (E. McCann & Ward, 2013), while focusing on "how regulatory practices and institutions achieve 'model' status" (Peck et al., 2009, p. 1096).

On the other hand, the variety of neoliberal-inspired policy and adaptations means that "critical commonsense" (Peck, 2004) around the meanings of neoliberalism are challenged and rewritten in the grammar of space-based analysis. Among these common ideas, we find the shifting role of the state: this has not been simply replaced by market forces in a gust of anti-regulation policies. Instead, as Foucault suggests (2008, p. 132), "Neoliberalism should not be identified with laissez-faire, but rather with permanent vigilance, activity, and intervention." In a neoliberal framework, while "the overall exercise of political power can be modelled on the principles of a market economy" (Foucault, 2008, p. 131), "markets themselves are not, never have been and cannot be spontaneously occurring and naturally self-regulating," as Peck (2004, p. 394) puts it.

For example, in our analysis of the European Innovation Partnership on Smart Cities and Communities (EIP-SCC) projects for 'citizen-focus' smart cities, Rob Kitchin and I (2019b) found a striking commonality of languages, propositions, and engagement practices in the designing, deployment, and implementation of the 'smart city'. This commonality suggests an active role of funding bodies in shaping the marketplace for 'smart city' initiatives. The EIP-SCC is an initiative of the European Commission, founded in 2011 with the aim of "bringing together cities, industry, SMEs, banks, research and other smart city actors," "boosting the development of smart technologies in cities," and "improving citizens' quality of life" by way of focusing on "the intersection of Energy, ICT and Transport."[3] The EIP-SCC is divided into six clusters[4] and a 'Marketplace'.[5] Each cluster is composed of projects and commitments intended as "*measurable* and concrete smart city engagements/actions from public and private partners" (emphasis added). According to its reports, there were 370 Commitments with over 4,000 partners from 31 countries in June 2014. For our research (Cardullo & Kitchin, 2019b), we undertook discourse analysis of policy documents and project descriptions of the 61 Commitments in the EIP-SCC 'citizen-focus' cluster.[6] We then interviewed a dozen stakeholders working on citizen engagement in a small sample of EIP-SCC flagship projects, asking questions about the different institutional arrangements and scales in the delivery of their 'smart city' projects, the time-line according to which their projects were prepared, funded, and institutionalized, and the actual existing spaces for citizen feedback and control within such projects.[7] Thus, supra-national funding agencies, such as the EIP-SCC, work as a mechanism of adjustments of opportunities and a platform which allocates resources, displays prepackaged solutions for various stakeholders, and favours exchanges within already determined boundaries of cooperation. This discursive homogeneity is further corralled by an active and transnational community of advisors from academia, the corporate sector, and the state: this suggests the flourishing of new hybrid configurations of state/market/academia relations, rather than the neoliberal state simply handing over functions and service to the private sector.

New forms of governance have emerged which displace decision-making as a democratic, contested, and inherently political process, instead, 'naturalising' economic and social relations as technical and operational matters, a domain covered by specific forms of expertise to which smart technology can offer an array of solutions (see Kitchin, 2015; Swyngedouw, 2011). As we will see more in detail in the following chapters, the 'smart city' comes with a healthy pedigree of algorithm-enabled governance practices which, rather than democratising society and participation, seem to widen the gap between citizens and decision makers. In the next section, I will look at the panorama of 'citizen-centric' smart city initiatives while suggesting an analytical framework which recalls the main characteristics of neoliberal urbanism, as its policy trajectory and interurban implementation travel across specific sites (in my focus, Europe).

(Neoliberal) Smart urbanism

Although neoliberalism comes in many different forms and its ideological underpinnings have been translated into policy in many ways, I group the main features of the technological turn in the neoliberalisation process of cities around five broad tenets: (1) smart urbanism strives to capture public assets and services by offering technological solutions to urban problems (*technological solutionism*); (2) steering citizens' behaviour *(nudging)*; (3) often, by way of using financialisation of services and short-term 'disruptive' solutions (*marketisation*); (4) moreover, smart urbanism fosters local economic growth while attracting foreign direct investment, for instance, via Living Labs and incubators for start-ups or digital hubs (*testing, scaling and replication*); in addition, (5) neoliberal urbanism drives real-estate investment through processes of financialisation of home and the transformation of urban space to mere exchange value (*smart enclaves and creative classes*).

Technological solutionism

Neoliberalism involves the systematic curtailing of public spending, competences, and funding available to cities to organise and update city services and functions. In the last 30 years, the neoliberal offensive to the public sector resources has been relentless and ubiquitous, leading often to austerity politics which have severely cut city budgets for employment and skills upgrade and for the maintenance of urban infrastructures (e.g., Peck, 2012). As a consequence, austerity politics have pressured cities and other public institutions to privatise and outsource public provisions, for instance under the 'smart city' agenda. Faced with budget constraints, it appears 'rational' for city administrations to draw on the competencies held within industry in order to formulate 'smart city' policy and to deliver tech-led services through public-private partnerships, leasing, deregulation and market competition, and privatization (Shelton et al., 2015). Cities, it is argued, are behind the technology curve with respect to state-of-the-art ideas and management systems: they lack the core skills, knowledges, resources, and capacities to address pressing urban issues and maintain critical services and infrastructures, which are becoming more socially and technically complex and require multi-tiered specialist interventions (Kitchin, Coletta, Evans, et al., 2017). Within this mindset, the place of the public sector is to act as broker, rather than service provider, with 'smart city' units seeking to source initial expertise and build partnerships with industry and academy.

Thus, the 'smart city' largely takes a technological solutionist approach to solving urban issues (Kitchin, 2014b). That is, there is a presumption that all aspects of city functioning and life can be mediated, treated, or optimized through technological solutions. All that is required to solve deep-seated and long-standing social issues such as congestion, energy consumption, emergency management of events, sub-optimal behaviour, crime, and the very functioning of the democratic process are data-driven software solutions. As we will see in the next chapter, the shift towards technological solutionism has a great impact on the concept of democracy itself, where inherently agonistic, potentially contested, and politically grounded ideas and decisions leave room for discourses based on algorithmic rationality, automatic responses, and technical efficiency. As Fuller writes (2017, p. 219), this technocratic approach shifts "the question of what is just to the question of what is technically best," ultimately operating, in Foucauldian terms, a shift "from politics to equipment."

We can say that the way in which the 'smart city' concept has been mobilised gives plenty of ground for a technological solutionist approach. For critics, in fact, the dominant 'smart city' discourse has been rooted in ecological preoccupations as "consensually agreed metaphors" (Swyngedouw, 2016) or "stage-managed consensus" (MacLeod, 2011). In China, for instance, "eco-cities" have been framed as the blueprint for smart urbanism, thus merging techno-scientific solutionism and ecological preoccupations for sustainable urbanisation (Caprotti, 2014). Similarly, the European Commission has set the key objective for 'smart cities' as the "transition towards a low carbon and resource efficient economy" – where urban EU populations are said to consume "70% of our energy" (European Commission, 2016, p. 105). As this narrative suggests, the implementation of 'smart cities' is a shared and urgent paradigm for our planet as it becomes ever-more urbanised. In other words, the 'smart city' is fundamentally ideological in its premises, although it carefully removes politics from the equation and substitutes this with forms of algorithmic governance and largely agreed-upon objectives.

In our mapping of the citizen-centric programmes (Commitments) submitted under the EIP-SCC funding scheme from 2015 to 2017, we found a large number of city 'interfaces' working through apps and dashboards, generally utilising real-time data (public or not), aimed at solving urban issues (Cardullo & Kitchin, 2019b). By following the 'trajectory' (Ward, 2010) that public policy takes in these many 'smart city' Commitments, we can get a holistic picture of the symbolic and operational landscape of their conceptualisation, design, and implementation – although each case is different and there is no attempt here to evaluate these in detail. The focus on policy trajectory is important as "neoliberalisation denotes a direction rather than a destination," following "a zigzagging path of creative destruction" (Peck & Whiteside, 2017, p. 181): that is, the local implementation of neoliberal policies is variegated and contradictory, while responding and adapting to the steering of supra-national and inter-urban policy objectives, networks, and funding regimes.

With regards to the present discussion around technological solutionism, Commitment 148 (SmartAppCity), for example, promises a mobile application that integrates and presents all city services via a smartphone app aimed at "improving [citizens'] quality of life and generating wealth" (ported in conjunction with a "Geomarketing tool able to offer promotions and events to users who are close to their stores"). Another Commitment, The Green Network, promises to produce "quantum energy savings, improve urban rent, quality of life and attractiveness of districts" and improve "local and regional long term employment and growth" by refurbishing city districts with the "latest materials and technologies".[8] Clearly, such an approach is underpinned by an instrumental rationality that largely divorces an issue from its wider framing, context, and interdependencies, and the role of politics, governance, culture, and capital in shaping urban relations. Moreover, the operationalisation of these solutions are tracked and evaluated on a narrow range of measures. As a project manager working on one of these 'citizen-centric' initiatives lamented: "Project leaders and the council are all . . . like 'oh this is really important', but then all the meetings come back to: 'What are our deliverables? What are our *measurable* outputs? How do we achieve these *measurable* outputs?' Everything becomes about a spreadsheet at the end of the day" [my interview, emphasis added]. This passage is key in understanding the collusion of neoliberal urbanism with 'smart city' ideals: what counts, here, is the achievement of *measurable* benchmarks which reduce the complexities of urban life to quantitative output and, ultimately, to short-term financial returns. The place of the public sector, devoid of resources and responsibilities, is then to challenge companies to offer solutions to a set of urban problems, to make resources available, to facilitate stakeholder engagement, and to manage contracts.

Nudging behaviours

Parallel to this emphasis on technological solutionism, we probed our dataset of 'citizen-focused' Commitments to observe the increasing trend of envisioning citizens as 'learners' with the aim of educating them as to how to best to use resources or adopt a certain behaviour (Cardullo & Kitchin, 2019b). This has led to a cottage industry of apps which seek to educate and change behaviour, steering and nudging people towards an efficient model of urban growth that will simultaneously improve "*their* quality of city life" (European Commission, 2016, p. 105, emphasis added).

Increasingly, public engagement and participation take the form of 'gamification', where social problems and issues are managed through the mobilisation of game elements in situations other than the ludic (Vanolo, 2018). For instance, Clicks and Links[9] is a company which promotes "behavioural change through gaming and virtual reality" within CITY-ZEN,[10] a project that aims to engage and educate citizens about energy-efficient infrastructures. In a similar vein, Commitment 6939 (Energy GOALS) wants to deliver an "empowering game" aimed at 8- to 14-year-old children to support behavioural change leading to an energy reduction in social housing. Commitment 7422 (Cooperative Gaming) offers a game on energy efficiency and the use of renewable energy between neighbourhoods within a metropolitan region and between different EU cities. Commitment 7788 (Mondragon commitment) also advocates the use of smart platforms and gaming to foster "citizens behavioural change" for energy saving purposes and, in addition, offers the possibility for service providers "to gather a quick picture of *their* current sentiment." As Vanolo (2018) suggests, with the insertion of metrics of rankings, scores, badges, levels, rewards, and virtual currencies in apps and websites, the gamification effect too works towards colluding neoliberal urbanism with 'smart city' ideals, through "the logics of competition, individualism, rewards and responsibilisation of the self" (2018, p. 324). In other words, gamification is a specific form of nudging operated through elements of gaming.

There is a lot of talking about nudging as a way of fixing city problems via technological solutions in the array of documents, websites, and papers that support the 'smart city' vision. The bottom line of such reasoning is that technology is available or already in place; it is now up to its users to change their behaviour in order for this to produce the foreseen effects. There is no guarantee – and indeed there is scarce evidence – that such change of behaviour happens as the result of the technology adopted. As a project leader I interviewed recently admitted: "The project as a whole can run because you've said to the funders, in order to get your funding, 'these are the things that we are going to achieve' [including people's change of behaviour]; but sometimes you don't know that." Again, rather than engaging the direct interested people in a conversation about the future of their city, my impression is that behavioural change is much more about selling a message and manipulating outcomes. While one city official I interviewed said she was seeing the "already changing behaviour" of her fellow citizens recruited in a smart meters pilot for reducing electricity consumption, other interviewees expressed deep concerns around the suitability of smart meters as indicators for a change of behaviour: "We have talked quite a lot about it [change of behaviour] and how we measure that. We need to look at the quantitative data that we might get from smart meters but we want to understand the everyday lives of some people we are working with."

To repeat, there is no intention here to evaluate each project or initiative or Commitment – such a thing would require a much deeper and longitudinal engagement with each different context. More focused case studies will need to be carried out after projects have been delivered to get a true insight into how people understand and act with respect to, for instance, energy consumption. However, we looked at the pattern in the design and discursive construction of a fairly large group of projects in order to detect their entanglement with neoliberal urbanism as both an ideology and a policy direction (Cardullo & Kitchin, 2019b). So far, there has been no evidence that a change of behaviour has actually affected the community in question in any meaningful way, and for the common good. Rather, the framework deployed is clearly rooted in technological solutionism and in a notion of individual citizenship which is instrumental to private provision of public services.

Marketisation and short-term 'disruptive' solutions

The climate of market-led and technology-enabled solutions to urban problems and the continuous cuts in personnel, training, and resources the public sector has been subject to because of austerity measures have determined a withdrawal of cities from the cutting-edge research and development of technological innovation and from upgrading of their own infrastructures. This has been particularly evident and dramatic during the explosion of the coronavirus health emergency. Thus, the high-tech sector has been able to actively disrupt existing public and private services and infrastructures, and their regulation and labour relations, by providing new tech-enabled platforms – for example, Uber and Airbnb challenging traditional taxi and short-term accommodation markets. This shift in perspective is recognised explicitly in public funding calls for 'smart cities': one of the main forms of impact for initiatives seeking funding is to attract significant private investment in the delivery of public services. So we learn that a "good impact" would be to show a reduction of "the technical and financial risks in order to give confidence to investors for investing in large scale replication" (European Commission, 2016, p. 111), so that eventually "private capital can take over further investments at low technical and financial risks" (ibid., p. 108). In other words, there is an offer for the socialisation of risks in exchange for the privatisation of services and, eventually, profits.

At times, the slippage between citizens, users, and consumers is evident: the H2020-SCC call suggests as a meaningful impact that "the active participation of *consumers* must be demonstrated" (European Commission, 2016, p. 107, emphasis added). In contrast, we find only one mention of 'citizens' in the impact section, with the goal of making "local energy system more secure, more stable and cheaper for the citizens and public authorities" (ibid., p. 111). But what kind of 'citizen' is implied here? The installation of smart meters in their own home, or the incorporation of renewable energy sources, hardly gives citizens an "active participation" or a say in the running of the electricity company or grid. Rather, the citizen is a consumer, in a marketplace of privatised utility provisions, and a product (as data). The citizen is useful to the extent to which she or he can produce revenue and valuable data for the company and for the deliverable of the Commitment itself.

While cities are left to struggle with increasing demands and less resources, companies try to capture private infrastructures and services. They then seek to extract value through minimizing maintenance and long-term investment and charging the highest bearable price depending on a user's ability to pay. Indeed, while visiting the 2017 & 2019 Smart City Expo World Congress (SCEWC)[11] held annually in Barcelona, and talking to many private-sector representatives, engineers, and CEOs, it appeared clear that private companies are ultimately, if not exclusively, relying on public money to expand their smart initiatives (Cardullo & Kitchin, 2019b). Mayors and city officials are seen overwhelmingly as customers, the smart interlocutors who are willing to invest in a problem-solving technology. To repeat, two complementary processes work to enable such a shift. First, cities struggling with tight budgets become increasingly reliant on competitive funding from supra-national bodies in order to implement technologies and services. Second, austerity is driving city administrations towards outsourcing and procurement of smart solutions that are purported as necessary to a city's own competitiveness (best practices *among* themselves) and as energy/labour-saving (best practices *within* themselves).

Testing, scaling, and replication

Scaling and replication are two crucial and interconnected issues at the heart of the 'smart city' strategy at the European level. Scaling seeks to bring forth 'best' solutions and translate successful pilots into deliverables. This strategy uses prototype pilot studies and *in-situ* trials to produce market solutions that can be deployed elsewhere. In order to create confidence and a climate favourable to risk-taking investments, scaling aims "to test and validate the business model" [interview with project manager]. When funding for pilots ends, initiatives are vetted with respect to their sustainability with regards to the city and to "the industrial partners and the industrial stakeholders that are also involved in a project, so they can see how they can replicate this in other areas and do business" [interview with project manager]. That means new service provisions are evaluated through efficiency criteria which, in the neoliberal austerity framework, translates necessarily into savings (doing more with less) of both physical and human resources and in the introduction of payment schemes in the medium term.

Replication is the process of translating scaled technologies and policies in other locales and it is a key stage for the implementation of 'smart city' policies. While scaling seeks to demonstrate local application, replication seeks to demonstrate generalisation and mobility; that 'smart city' initiatives proven in one place can be deployed with similar results elsewhere. It is through this process that transferable technologies, models, or 'best practices', and their circulation are established (E. McCann & Ward, 2013). In the case of EIP-SCC, this occurs through the institution of the Marketplace: Lighthouse cities work together to pilot and scale initiatives before Follower cities seek to replicate their work with the aim of creating a feedback loop that can inform the initial deployment, as well as create a case for wider deployment. But replication presents a circular rationale. The Lighthouse status is "itself the product of discursive attribution" through which applicant cities have been awarded, and thus certified by the EC, as being 'outstanding' smart cities (Engelbert, van Zoonen, & Hirzalla, 2019). At the same time, Commitments are projects which endorse an *already specific* version of the 'smart city': for instance, Commitment 7388 (Ravenna Common Ground) advances a "device aimed at providing the community [with] a reinforcement in a smart perspective." Commitment 7283 (The Educating City) wants to develop "interoperable platforms and devices . . . to provide support to the objectives set up by EIP's Strategic Implementation Plan regarding citizens' involvement and their awareness." The circularity between smartness as the 'fix' to city problems and its spinning mechanism is here evident.

Smart enclaves and creative classes

Beyond making the city a market in and of itself, the neoliberal 'smart city' is an explicitly economic project, aiming to attract foreign direct investment and mobile creative elites, by way of fostering innovative indigenous start-up sectors or digital hubs. Cities around the world have created 'smart districts', designating an area of the city as a testbed for companies to pilot new technologies (Cardullo, Kitchin, & Di Feliciantonio, 2018; J. Evans, Karvonen, & Raven, 2016). In the UK, the Department of Business, Innovation and Skills has funded 'smart city' initiatives with the aim of positioning the UK as a leading exporter of 'smart city' consultancy and technologies (Taylor Buck & While, 2017). The digital economy is seen as a key sector for generating new employment and 'smart city' initiatives, a means to attract talented workers and facilitate economic activity, as well as being a new market opportunity. Thus, digital businesses need to locate in an ecosystem of suitable office buildings with high-quality technical systems and infrastructure, a strong concentration of business and support services, and a pool of suitable labour. One way to create these conditions is to regenerate an existing city area, one that occupies a central site near key transport links and other business services, repurposing or replacing existing buildings. Here, smart enclaves and digital hubs are seen as central to an ongoing process of 'modernisation' of the city, achieved by extending pioneering small-scale projects, design-focused Living Labs, and an entrepreneurial culture of open innovation to the overall organisation of urban space and living.

In the case of Dublin, Ireland, in the city centre there are two key sites of agglomeration, both of which are regeneration initiatives, redeveloping old, largely vacant or former industrial sites: The Digital Hub and Silicon Docks (Cardullo et al., 2018). In both cases, the primary focus is on growing the digital economy and regenerating the area into a vibrant economic zone. Other examples of smart enclaves include the 100 smart city developments in India (Datta, 2018), Masdar in United Arab Emirates (Cugurullo, 2018), Hudson Yards in New York City (Mattern, 2016), Milano Due in Italy (Di Feliciantonio, 2019), and Songdo in South Korea (Shin, Park, & Sonn, 2015).

An important feature of intra-city competition and urban entrepreneurialism is that they fit well with a speculative approach to housing, privatisation of space, and attraction of more affluent buyers: these are all well-documented characteristics of neoliberal urbanism, which conceives urban land via exchange value rather than use value (e.g., Kitchin et al., 2012). While there are some attempts to engage with local communities through Living Labs initiatives, these are largely tokenistic examples that play out good corporate social responsibility, as opposed to creating a 'smart city' from the bottom up. Rather than local communities fully benefiting from economic revitalisation, the creative classes are being drawn into these new digital hubs, displacing existing residents through soaring rental and property prices. As such, these areas can become key active sites of gentrification, where local authorities purposely seek gentrification as an ideal policy solution for urban change (Lawton & Punch, 2014).

In addition to urban-focused economic development, then, the 'smart city' has become a key component of property-led development. Here, 'smart' technologies are a central feature of new real-estate projects, operating as an attractor for investors and future residents, as well as providing a shopfront for those technologies for other prospective development sites.

Concluding remarks

The chapter has drawn on the research done by Rob Kitchin and myself (2017–2018) around the 'citizen-focused' framework deployed by the European Innovation Partnership for Smart Cities and Communities (EIP-SCC). It shows how, under a neoliberal and transnational framework, active forms of marketisation have been taking place, with assemblages of neoliberal governance able to form and move through diverse cities across Europe and at multiple scales. Our analysis suggests that the EIP-SCC works as a mechanism of adjustments of opportunities and a platform which allocates funding, displays prepackaged solutions for various stakeholders, and favours exchanges within already determined boundaries of cooperation (Cardullo & Kitchin, 2019b). Following the 'trajectory' (Ward, 2010) that forms of governance take in these many 'smart city' Commitments, we wanted to provide readers with a holistic picture of the symbolic and operational landscape of their conceptualisation, design, and adoption. While the local implementation of neoliberal policies is variegated and contradictory, these seem to respond and adapt to the steering of supra-national and inter-urban policy objectives, networks, and funding regimes. This configuration presents neoliberal urbanism as a 'sticky' ideology, rationality, and policy agenda that operates at the urban and intra-city scale at European and national levels. Through the work of a shared framework, such as the transnational funding schemes and the process by which projects are conceived, evaluated, and delivered, neoliberal ideals are transmitted and modelled on the dogmas of efficiency (saving scarce energy), sustainability (changing policy orientation in the long term), and freedom of choice (although instrumental to market imperatives).

In this chapter, I have highlighted five specific forms of governance through which the process of marketisation of the 'smart city' takes place: technical solutionism, nudging behaviour, financialisation and short-term 'disruptive' solutions of service provision, scaling/replication, and creation of smart enclaves. These forms of neoliberal governance are woven into the rhetoric of transnational 'citizen-focused' initiatives while, at the same time, actively circumscribing specific roles for the citizen. In other words, neoliberal urbanism works at a multi-scalar level, creating a paradox for the neoliberal subject. This is because neoliberalism shifts citizenship away from inalienable rights and the common good towards a conception rooted in individual autonomy and freedom of 'choice', and personal responsibilities and obligations (Ong, 2006; Vanolo, 2016). The onus to navigate and negotiate the provision of services and levels of access is then placed on the individual, based on their personal, social, political, and economic capital, and framed within 'commonsensical' constraints (such as the reduction of externalities brought about by global urbanisation). As Wendy Brown (2016, p. 4) suggests, "as neoliberal citizenship sets loose the individual to take care of itself, it also discursively binds the individual to the well-being of the whole."

The chapter has followed closely the way in which the 'citizen' is operationalised in many 'smart city' initiatives, attending to the journey of policy from the transnational context of the EC funding strategy to its interpretation, and eventual adaptation in places. An argument this book brings forth is that the increased reliance on big data analytics, city-sensing, and social-media interactions (activated within a framework of technological solutionism) might privilege real-time planning decisions over political discussion and agonistic processes of governance. As the next chapters will unpack in more detail, 'smart city' assemblages of governance have been promoting a model of participation that is rooted in pragmatic, instrumental, and paternalistic discourses and practices.

Notes

1 https://theintercept.com/2019/02/02/shoshana-zuboff-age-of-surveillance-capitalism/
2 http://progcity.maynoothuniversity.ie/resources/publications/
3 *https://eu-smartcities.eu/*
4 These are: 'citizen focus'; 'business models, finance and procurement'; 'integrated infrastructure and processes'; 'integrated planning, policy and regulations'; 'sustainable districts and built environment'; and 'sustainable urban mobility'. https://eu-smartcities.eu/clusters
5 The Marketplace aims at providing the platform (a "network of networks") through which cities and stakeholders can collaborate. The enterprise received initial funding of €18 million, which increased exponentially to €365 million only 2 years later, making it central to the EC policy goals of replication, emulation, and translation.
6 It is important to note that the EIP-SCC portal has undergone a complete overhaul since our initial research, making it difficult to locate the original details for each Commitment – although you can see a limited list at: https://eu-smartcities.eu/commitments. Much of the original Commitments are no longer publicly available and we can only refer readers to the list of Commitments archived by the 'Way Back Machine' – http://web.archive.org/web/20170416191724/https://eu-smartcities.eu/commitments. This is in itself troubling given the 'citizen-focus' cluster's supposed ethos of data openness, democratic governance, and public engagement.
7 Interviews were conducted in two European cities in October and November 2017.
8 http://ec.europa.eu/newsroom/dae/document.cfm?doc_id=4618
9 http://clicksandlinks.com/dvteam/city-zen/
10 www.cityzen-smartcity.eu/home/about-city-zen/calendar/
11 This is said to be "a leading platform of ideas, networking, experiences, and international business deals," which in 2019 featured 24,399 visitors and 1,010 exhibitors from 146 countries and more than 700 cities, "all committed to working together to build Cities Made of Dreams": www.smartcityexpo.com/en/. During our visit, I conducted 20 short and targeted interviews with city officers and corporate exhibitors.

3 Post-political governance and data ethics

In Chapter 1, we saw how the current neoliberal framework through which the 'smart city' ideal has been implemented is increasingly reliant on big data analytics, city-sensing, and social-media interactions. These are activated within a framework of 'technological solutionism' which privileges real-time algorithm-led planning decisions over political discussion and agonistic processes of governance. Chapter 2 discussed neoliberal smart urbanism, arguing how the 'smart city' is now a privileged gateway through which the neoliberalisation process of urban life takes place. Drawing on critical scholarship and the analysis of the 'citizen-focus' cluster of the European Commission initiative for smart cities and communities (EIP-SCC), the chapter further suggested that five forms of governance are at play within the current construction of the 'smart city': technological solutionism, nudging behaviour, financialisation and short-term 'disruptive' solutions of service provision, scaling/replication, and creation of smart enclaves.

The changing forms of governance in the 'smart city' and the new emphasis on ethics of data are going to be the objects of this chapter. This starts pulling together the different strings attached to discursive construction, actual implementation, and ethical implications of the neoliberal 'smart city' in relation to citizens. Following Deleuze (e.g., 2002), critics argue that algorithm-driven technological solutions are changing the forms of governmentality – the logics, rationalities, and techniques through which the art of government is possible. Computational processes are shifting governmentality from a disciplinary frame, enacted through direct surveillance and punishment, to capture and control, enacted through the modulation of affects, desires, and opinions, and by way of nudging within prescribed behaviours (e.g., Kitchin, Coletta, & McArdle, 2017; Sadowski & Pasquale, 2015). The context-dependent analysis that underscores this book highlights variety in the configuration and deployment of sociotechnical assemblages in enacting governmentality; for instance, "through modes of surveillance and discipline or capture and control; through systems that are 'black-boxed' or transparent; through regulatory techniques such as coercion, co-option, self-disciplining, punish, modulation, intervention, mediation, coordination, direction, and optimization" (Kitchin, Coletta, & McArdle, 2017, p. 16). Although still framed within the wider political, social, economic and legal landscape, forms of power and control are in fact mutable, driven by differing value systems and dependent on local and national institutional politics and policies and practices of deployment – for example, whether these depend on technologies of second- order cybernetics, fostering automatic and autonomous forms of feedback with human-off-the-loop configurations, or whether they still depend on humans-in/on-the-loop interactions (Coletta & Kitchin, 2016).

In addition, this change in governmentality happens within an ongoing neoliberal reorganisation of life and society: that is, the shift to control has also been accompanied by a broad change in the social contract between the state and citizens, from public infrastructures built for the common good to a form of corporate contract through which city services are delivered via public-private partnerships or private provision only (Sadowski & Pasquale, 2015). Advocates of the 'smart city' suggest that cities can be made more efficient through aligning and regulating infrastructures, services, and people to reach efficient and sustainable outcomes. In order to do that, increasingly, the emphasis has been put on the response by the publics (e.g., via behavioural change, feedback mechanisms, or citizen-sensing). What is demanded here are rational individuals/consumers (who respond to choices made for them) and compliant tech companies: the former are to be nudged and directed at distance, but within the ethical and legal constraints of the latter.

We can think of governmentality also as a reflexive discourse and cultural construction: the 'smart city' in itself becomes a goal for urban development projects and an assemblage of different urban imaginaries: as seen in Chapters 1 and 2, various ideological dispositions have blended technological advances in computing and informational systems with neoliberal policies for managing urban growth and controlling urban life. Ultimately, such dispositions feed into a sort of 'smart-mentality': this is a discursive arrangement which involves a "new geometry of power relations requiring the production and circulation of knowledge, rationalities, subjectivities and moralities suited to the management of the smart city project" (Vanolo, 2014, p. 894). This can be understood from both the perspective of cities (becoming 'smart' in order not to lag behind) and their citizens (who are therefore invested in a sort of moral obligation to adhere to the 'smart city' ideal, to become 'smart citizens'). As a consequence, Vanolo suggests, "smartness is becoming a field of social control that makes intrusion in a person's private life quite natural" (ibid.).

The present chapter suggests that a change in the way in which power imagines its subjects necessarily modifies governance, which is the actual practice through which policy is executed and decisions are formulated, as well as a communication mechanism between local government and residents. Since *governmentality* is shifting towards imagining citizens as subjects of constant modulation through software-mediated systems (which capture their movements, preferences, and even feelings), a more democratic model of city *governance* is becoming more urgent. This would imply, for instance, forms of citizens' inclusion and participation, open data centres and transparent AI, and e-government with effective accountability, community consultation, and local empowerment.

Ethics and the 'smart city'

As argued thus far, the current 'smart city' configurations mostly operate for private profit goals or for the government at a distance from the population. Such a configuration has attracted a formidable battery of critiques and concerns around the deployment of smart technologies and the unwanted effects they generate. Following Kitchin (2016b), some of these risks involve privacy, datafication, dataveillance and geosurveillance, profiling, social sorting, anticipatory governance, and nudging, all forms of governance that have significant consequences for how citizens are conceived and treated. Moreover, critics have pointed out the risks connected to technological mass unemployment (Fuchs, 2018), discrimination against minorities (Eubanks, 2018), disempowerment of less 'smart' people (Smart, forthcoming), and contamination of political processes (Gerbaudo, 2019). The present section starts addressing the ethical debate around the 'smart city', by asking: How do these modes of governance work in practice? What kind of unwanted effects does 'government at a distance' have on citizens? And, are the now sprawling ethics committees enough to limit the negative effects of smart technologies in cities, or able to influence the way in which we build and deploy them? Under a technological landscape of 'spatialised intelligence' (Picon, 2015), for instance, large parts of road networks are *de facto* monitored by inductive loops, traffic cameras, and automatic number plate recognition cameras (e.g., Coletta & Kitchin, 2016). In a number of cities, sensor networks have been deployed across street infrastructure – such as bins,[1] lampposts,[2] and benches[3] – able to capture and track phone identifiers such as MAC addresses. The same technology is used within malls to track shoppers in order to capture metadata in the forms of basic demographic information such as age and gender; or, increasingly, deployed with CCTV and facial and emotion recognition algorithms in order to have market research data on shoppers' behaviour or preferences, and thus predict buying trends.[4] Similarly, some cities have installed public Wi-Fi mesh networks which can capture and track the IDs of devices that access the network for various forms of spatialised analysis based on big data, for example in 'Digital Twins' simulations or for creating 'urban vitality' indexes (Kim, 2018). Moreover, most buildings and public transport systems monitor smart cards used to access them, and there are an increasing number of sensors for all sorts of monitoring (temperature, light, CO^2, and even room bookings, desk space, and diary synchronisation).[5] Notoriously, phones communicate their location to telecommunications providers continuously, either through the cell masts they connect to, or the sending of GPS coordinates, or their connections to Wi-Fi hotspots, or by enabling apps-specific features and Bluetooth technology (e.g., Greenfield, 2017).

This extraordinary technological push to inherently surveil, track, and control everything extends to other more intimate spaces, such as the home environment and the body. Thanks to their interconnected sensors abilities, Internet of Things (IoT) devices extend their monitoring capabilities well beyond the public sphere, to colonise private domains and leisure time: smart meters, voice-operated domestic assistants, digital implants and bracelets, and even remotely controlled dildos.[6] As seen in previous chapters, networked sensors and real-time big data streams complement the paradigm of the neoliberal subject grounded in individual responsibility: for instance, the growing practice of counting one's daily steps or measuring dietary targets are ways of analysing one's own data and then recalibrating one's behaviour to them (Davies, 2015). Apparently free from legal interfaces and physical market boundaries, this essentially active and almost entrepreneurial character (the 'smart citizen') is in constant search for affirmation and improvement (see Brown, 2016). With the coupling of personal and environmental sensor data, 'smart living' can lead to a gamification effect which determines notions of 'good' or 'bad' citizen/user through disciplinary dispositives of ordering or ranking (Vanolo, 2018). Han (2017, online) calls it "smartpolitics," arguing that the politics of discipline and punishment are being replaced by exploitation of the psychic realm: "Instead of forbidding and depriving, [neoliberalism] works through pleasing and fulfilling."

It is very likely that such enormous data gathering has profound implications for individual and collective rights to privacy, freedom of expression, and related dimensions of everyday life, from identity theft to political censorship. According to Han (2017), the neoliberal subject is not a 'labourer' any more, but a 'project' of governmentality: while the *onus* to navigate and negotiate the provision of services and levels of access is still placed on the individual (based on their personal, social, political, and economic capital), the framing of this intervention rests on commonsensical constraints (e.g., the reduction of externalities brought by global urbanisation: pollution, congestion, depletion of natural resources, climate change, and so on). This strangely echoes with the policy nudging of contact tracing apps introduced very recently in order to fight coronavirus spread, to which I dedicate further thoughts in the last chapter. The following two examples from China bring into the fold the various assemblages that form governmentality in the 'smart city', showing exactly the shift towards forms of 'governance at a distance' that surveil and control rather than directly disciplining and punishing.[7] It is important to stress that China is only one case study, although of relevance, in the evolving geopolitics of 'smart cities' deployment, showing how policy and marketing goals intersect in unforeseeable, characteristic, and inherently dystopian ways. As Vincent Mosco suggests (2019), the 'smart city' agenda only accelerates the "militarisation of the world's cities," which has indeed a long history of commercial and political connections between the military state apparatus and the surveillance capitalism of our times; this dating back to the beginning of the 'network of networks', and further enabling near complete police surveillance of most public and private places.

China is a particularly relevant case in point for the study of algorithm governance because it brings together the paternalistic gaze of the totalitarian state with the tracking practices and capabilities of corporate high-tech capitalism. These two forms of data-veillance and control are thus enabled within a framework of 'state capitalism', where the state sets the scene for private companies eager to extract huge amounts of profitable data from the burgeoning Chinese markets for consumers' goods and services. The Chinese government has in fact invested "heavily in Next Internet[8] technologies going as far as to integrate them into its five-year plans" which massively benefited private companies like Alibaba,

Baidu, Huawei, and Tencent (Mosco, 2019, p. 213). These companies[9] have been appointed by the Chinese government as first members of an artificial intelligence national team in order to lead on its future developments (Ding, 2018). The effect on the politics around ethics is painfully evident: imagine Toyota and Skoda being co-opted by a totalitarian state in order to write the Traffic Regulations!

The so-called "Sharp Eyes" is the government project that wants to ensure there will be one camera for every three people in China (a CCTV system which claims already to cover 100% of the surface of Beijing). The system will be enabled with algorithmic intelligence for face-recognition abilities.[10] The rhetoric of public security is now imbued with the claim to efficiently managing the urban crowd for the optimisation of flows and resources, where the 'smart city' has become synonymous with sustainable urbanisation or 'eco-cities' (see Caprotti, 2014). At the same time as directly surveilling through the 'Sharp Eye' programme, China is experimenting with a Social Credit System (to be fully implemented by 2020). This algorithm system rewards people for repaying debts promptly and factors in "court judgments, criminal records, academic dishonesty, jaywalking, moving violations, and failing to pay transit fares" (Pasquale, 2019). The 'good citizens' are incentivised to share their social credit score, thus increasing the gamification effect on governmentality (see Vanolo, 2018). Since notions of what is 'good' belong to the sphere of ethics and to the formulation of what 'value' is for a determinate society, Pasquale (2019) calls these systemic transformations, which determine social benefits for citizens, the "quantitative governance of culture." Thus, the ultimate aim of such a governance mechanism (based on personal and often sensitive data) is to produce a social index of the 'good citizen' congruent and compliant with the *status quo*. Online activities such as forms of citizen journalism have, for instance, a generally negative impact on the civic score, which rather nudges people towards consent.

These emerging practices of data-led predictive governance include policing, with analytics used to assess likely future behaviours or to anticipate the location of future crimes, so that police forces monitor the communications of known activists to try and anticipate and control social unrest (Paasche, 2013). This has led to a form of data determinism in which individuals are not simply profiled, judged, and treated on the basis of what they have done, but on the prediction of what they might do in the future. As Kitchin writes (2016b), a person's data shadow does more than follow them; it precedes them. But there is a growing concern that China is using smart technologies, such as face recognition, to also police ethnic minorities, for instance, the Muslim Uighurs who are actively monitored and persecuted. According to an investigation by the *NY Times*, this is the first known example of a government intentionally using artificial intelligence for racial profiling, a sort of 'automated racism',[11] although others have pointed to the institutional racist policing in the US via algorithm-enabled decision (Eubanks, 2018) and both IBM and Amazon have announced the deferral of face recognition projects in the wake of anti-racist movement across the US.[12] Thus, algorithmic governance models like China's massive investments in smart initiative "rebrands repression as rational nudging,"[13] by way of practising predictive privacy harms, social sorting, and redlining of population cohorts.

As for the examples of China, the changing modes through which government understands its logics and functions in relation to its citizens (governmentality) have pushed technological innovation towards hybrid forms of governance which *simultaneously* capture, surveil, and control. These are driven by big data and enabled by the artificial intelligence which understands them as evidence-based governance and real-time feedback. There is a growing preoccupation within critical scholarship that urban initiatives driven by technology companies fail to understand that "city-making is always, simultaneously, an enactment of city-knowing – which cannot be reduced to computation" (Mattern, 2017). We have framed this paradox as the epistemological bias of urban science which *both* knows the city through big data *and* enacts solutions accordingly via ad-hoc system thinking and stochastic modelling. Moreover, when smart devices are directly supplied by private hi-tech firms or on behalf of the state, there can be issues of ownership of the data generated, control over their processing, and accountability over the many actants that institutionalise the different technological assemblages.

In this sense, the 'smart city' agenda would need to rethink notions of behaviour change to recognise that, for instance, urban mobility is a form of 'spatialised intelligence' (Picon, 2015) to the extent that it is constitutive of social practices rather than dry statistics only (e.g., Kim, 2018). Mobility practices are in fact linked to the geographies of everyday life and the emotional and corporeal connections we have through movement. For instance, our intimate sense of place is also necessarily gendered, in that women's perception of place safety can be very different from that of their male peers.

An ethics of compliance?

As suggested thus far, the coming together of different technological layers – which inherently surveil, control, and track movements, spending habits, and civic behaviour – is of particular concern for critics of 'smart cities' and AI systems. This preoccupation has opened a much needed debate in recent years on how to articulate an ethical 'smart city' that safeguards the bundle of (digital) citizenship rights while favouring an emancipatory policy towards the 'right to the smart city' (Cardullo et al., 2019). While the latter is going to be widely debated in the second part of the book, the former presents itself as a short-term accomplishment of the 'good city'. Following a recent (2019) conversation with Rob Kitchin, I want to distinguish here between these two frameworks that seek to address unbalance or inequality in 'smart city' technologies: the technical and compliance model *vis-a-vis* the normative and ideological one.[14] Both stances have an inherently humanist and progressive framing, in that they want to ameliorate or change the current *status quo* with regards to citizens' use of these technologies. They both attempt to 'do good' with innovations or help citizens to move up the 'scaffold of participation' (Cardullo & Kitchin, 2019a), or they might even foreground utopian social aspirations.

However, the first model presents an increasing preoccupation with ethics as a guidance to 'do good' by way of correcting glitches in the 'smart city' systems (at least, those that become evident) without changing the *status quo* or the fundamental structures of society (e.g., the market-led impulse for creating the 'smart city'). It follows the tradition of Utilitarism and Deontological Ethics, that is, in procedural and compliant schemes of behavioural practices. Ethicists in residence, ethics committees, and a more widespread culture of supervision and compliance seem to be the logical outcome and the technical answer to 'issues' such as the collection, handling, and processing of data and to 'barriers' towards the adoption of these technologies. At best, they can help reframe a product or a service as 'ethical' and thus exploit a competitive advantage in the industry of smart technology production (e.g., EU-compliant drones or an app that is certified with a 'privacy-by-design' logo).

The second model, the normative and ideological one, would instead look at the power relations that make such 'issues' possible in the first place, for example, structural inequalities in society, power relations, and effective forms of participation. This model is holistic in its approach, following a tradition proper of Aristotle's *poesis*. It would hold, for instance, that the correction of biases in the various algorithms or apps that regulate everyday life does not eradicate gender and race inequalities, which are deeply entrenched issues in society. If bias is a social problem, seeking to solve it within the logic of automation is always going to be inadequate, because this only reflects pre-existing biases in society.

For critics (e.g., Kitchin, 2019a; Metzinger, 2019; Wagner, 2018; Yeung, Howes, & Pogrebna, 2019), the compliance model raises suspicion of 'ethics washing', which aims at pre-empting or avoiding more serious issues of social justice and effective participation from below. 'Ethics washing' can be thought as a way of organising and cultivating ethical debates to distract the public from preoccupations related to new technologies and, thus, to prevent more effective regulation and policy-making. Once the algorithmic process leading to life-changing decisions (e.g., penal conviction or mortgage lending) has been made more fair, do we still need to think about inequality of access to the housing market (e.g., financialisation of housing and gentrification)? Or about predictive policing and governance based on race and gender (e.g., institutional racism in the police force or structural gender biases in the employment market for the IT industry)? While the ideological and normative 'model' requires some serious imagination which aims at tackling structures of domination, the compliance and procedural framework risks simply reproducing the *status quo* (D'Ignazio & Klein, 2020). 'Ethics washing' is spin off through window-dressing activities that bring together a coalition of 'smart city' advocates and an academic community of 'experts', and usually resolves to piloting initiatives that hardly tackle the issues at stake (see Kitchin, Coletta, Evans, et al., 2017).

At present, we can count about 70 ethical frameworks, guidelines, and principles on AI, robotics, and data (Floridi, 2019).[15] On the one hand, this can be a symptom of how urgent the topic is becoming. On the other hand, though, it is also a sign of general confusion, compelling compliance to existing or likely forthcoming legislation, and corporate opportunism. This proliferation of ethical frameworks highlights the 'ethics washing' strategy implied above: that is, pre-empting more serious issues and responsibilities. At the same time, 'ethics washing' indicates that companies are shopping around for principles and values (Wagner, 2018), in the sense that they try to retrofit their data practice to one of the existing frameworks (or even create a new one which fits their current practice), rather than putting in place practices that follow a widely accepted and consistent ethical framework.

In trying to reduce this 'noise' around ethics of AI, robots, algorithms, and data handling, in 2018 the European Commission appointed 52 experts to a new High-Level Expert Group on Artificial Intelligence (AI HLEG) who delivered their own framework and guidelines on ethics for algorithm-driven and deep learning processes.[16] It is worth remembering that AI is an umbrella term for a growing number of technological assemblages that are automated and mostly autonomous, and based on algorithms in order to deliver computational processing of big data in (almost) real time. The applications of AI are primarily for the exploitation of data for operations of industrial processes, logistics, predictive policing and governance, driving of autonomous machines, and deep learning computation. Although the European Union is the first political entity to deliver ethical guidelines on such a new and controversial domain, and their legal anchoring is rooted in European fundamental values (such as 'Human agency and oversight', 'Accountability' and 'Transparency' of AI), the commission has been accused of 'ethics washing'. According to its critics (Metzinger, 2019), the composition of the AI HLEG actually reflects the heavy weight of representatives from the industry.[17] As a consequence, in the final draft of the AI manifesto, the commission turned to more sober 'critical concerns' rather than the proposed 'red lines' of AI implementation. The latter are non-negotiable ethical principles determining what should *not be done* with AI in Europe – such as the use of lethal autonomous weapon systems, the AI-supported assessment of citizens by the state (social scoring) and, in principle, the uses of AI that people can no longer understand and control. The former, instead, leaves the door open to interpretations and political swings; in this respect, it is a form of 'ethics washing'. In sum, the oft-time advertised 'fixes' to artificial intelligence preclude *de facto* any other option, such as the possibility to never use such systems at all.

In the corporate field, too, there is evidence that, despite promises of corporate social responsibility and concern for users, the core lobbyists and lawyers of Big Tech are crushing any effort to protect citizens from privacy invasion and data misuse. For instance, Amazon and Google have contested vehemently a bill from the state of Illinois, the so-called 'Keep Internet Devices Safe Act',[18] which suggests a company is not allowed to turn on a microphone remotely without the owner's permission.

Moreover, ethics boards and committees have a tough time in making their view count in corporate and city environments that do not have a culture of openness or a political drive for change. At times, the compliance status is as far as one can get in terms of governance of smart technology. This is because of the absence of long-term goals or political insights, and because there is generally a sense that the 'smart city' is the unavoidable format, which we can tweak within reason (usually the reason of the market) and within the available legal framework (which is paradoxically said as being 'inadequate' to face the challenges posed by smart technologies). For instance, Google has been under the spotlight recently for first introducing an ethical committee on 'responsible development of AI', which was supposed to meet up only four times a year and without clear decisional power, and soon after for being unable to justify its composition to its own employees and to the wider public due to ethically compromised appointees on the board.[19] In this regard, a positive example from ethical compliance practices seems to be the Dutch ING Bank, one of the 15 largest in the world. Constantly praised in the press and winner of numerous awards from independent sources, the bank has a Data Ethics Council that meets once a month and apparently "really has the mandate to reject project proposals."[20] The scope of this example is not to evaluate ING group as being a more or less 'ethical' company,[21] but to show that sometimes ethical boards might actually work in a way that gives them veto power over unethical projects or decisions.

Indeed, according to the annual Corporate Accountability Index 2019[22] – which ranks the world's most powerful Internet, mobile ecosystem, and telecommunications companies according to their commitments and policies based on international human rights standards – the gap between corporate governance and 'good practices' around privacy and digital rights is persistent, but largely unaddressed by regulators of the industry. In other words, despite having some mechanisms in place (e.g., guidelines, committees, procedures), Big Tech companies generally fail to adopt these across their operations or towards their stakeholders, showcasing a lack of internal accountability mechanisms, board and corporate-level oversight, risk assessment and ethical audits, and employee training and grievance mechanisms. Similarly, the Ranking Digital Rights (RDR) Index represents the extent to which companies are meeting minimum standards on their *publicly disclosed* commitments and policies around digital rights; that is, it evaluates how transparent to scrutiny and compliant to current legislation these companies are. With only a few companies going above 50 percent in the Index, out of 24 evaluated,[23] some concerns are more than justified. This finding needs to be put in context, though, by considering that Big Tech represents around USD 5 trillion of combined market capitalisation and their products are used by the majority of the world's connected population (estimates to 4.3 billion).[24] In this regard, the report's recommendation that "companies must take more affirmative steps to respect users' rights" (ibid.) might appear awkwardly mild to more critical readers.

The above discussion on ethics moves within the muddy waters of corporate social responsibility and its implementation in practice: one obvious reason for concern is that this form of responsibility has to be *always already* compliant with corporate goals while also instrumental to the logic of the market. Drawing on the Foucauldian notion of subjectification, Banerjee (2012) argues that organizational and institutional discourses shape and govern individual managers, who become constituted as a subject through their relationship with the firm. Thus, from a critical perspective, "corporate social responsibility becomes an ideological movement designed to consolidate the power of large corporations" (2012, p. 8). There is a limited scope for management to seriously consider ethical issues related to AI, data, and technological adoption in everyday practices. Rather, if it were to make any changes, ethics in the corporate field would require a more holistic approach in the actual configuration of production, adoption, and uses of smart technologies; for instance, by having on the boards representatives from workers' organisations, environmental groups, city representatives, and civic society at large; that is, by enabling multiple and agonistic perspectives other than only the managerial and corporate one.

In order for ethics to become a normative issue, then, I would suggest we need more than ethics. We need history, sociology, psychology, political science, economics, and law, in order to integrate research and teaching around technology with the humanities and social sciences, as well as with political activism. More poignantly, the introduction of ethics and a politics of emotions in STEM-oriented studies and practices would represent a move forward for data science educators; these could benefit from partnerships with media, art, and design educators, whose fields are built on experimenting with ethical and equality questions (D'Ignazio & Klein, 2020, Chapter 8).

The post-political unfolded

As mentioned above, the 'smart city' discourse has been justifying a "largely depoliticized ideological rubric" (Brenner & Schmid, 2015, p. 158). That is, it has been built exploiting "consensually agreed metaphors" (Swyngedouw, 2016) offering an ideological framework which fosters technological solutions for deep-seated and long-standing societal challenges and urban externalities such as congestion, pollution, crime, unemployment, service delivery, and infrastructure updating, just to mention a few. This growing trend draws on the belief that technology can have specific effects, such as change of behaviour in their users and improvement of the quality of life (*technological determinism*), and can fix urban problems through, for instance, apps or sensors (*technological solutionism*) (e.g., Kitchin, 2015; Morozov, 2013).

Therefore, the 'smart city' has increasingly become a generic and easily agreed target, by-passing debates which foster different ideas and positions, and thus shifting towards post-political practices of governance (Swyngedouw, 2011). For instance, there may be a generic agreement in the public opinion that politicians are corrupt, but it is another thing to believe the solution to this lies in an app powered by swaths of *objective* data and processed by *unbiased* AI. Equally, there might be a consensus on the externalities produced by rapid urbanisation (usually congestion and pollution), but there is less agreement on the meanings of sustainability (environmental efficiency or social ecology?) and on the solutions to this (e.g., powering cars with cleaner energy or aiming to reduce private transportation overall?) (March, 2018). This series of paradoxes appear evident when sustainability becomes the rationale for the 'smart city', since the latter is built around neoliberal logics which push for urban growth while moving citizen involvement away from the decisional processes governing the life of their city. Moreover, in a post-political and techno-solutionist framework, a series of factors influence the lack of accountability from service providers: 'smart city' initiatives are often delivered by what Swyngedouw (2005) calls "beyond-the-state" or "hybrid configurations." A transparent and democratic process is further dissipated by the recent proliferation of flexible and decentred models of urban governance and new administrative units, quangos, public-private agencies and 'experts' (advocacy coalitions made of middle management, external contractors, data analysts, chief innovation officers) which co-constitute the splintering of infrastructural provision (Kitchin, Coletta, Evans, et al., 2017). Indeed, many 'smart city' initiatives and projects are implemented by agencies with little, if any, political oversight, let alone citizens' involvement and participation.

Advocating an 'intelligent city' beyond the technological infrastructures that make it 'smart', Sartori (2015) suggests the 'smart citizen' is a social actor who uses digital technologies in order to interact with *both* the communities she belongs to *and* city governance. In multiple ways, this 'smart citizen' contributes to the social and civic infrastructures of her city. Therefore, "cities infrastructural hardware has to be accompanied by a sophisticated and open social software, whose code ought to be accessible and re-programmable" (Sartori, 2015, p. 943, my translation). This is the profile for an active citizen, in the sense that she engages with and contributes to the construction of the urban social milieu she is embedded within: more and more often, we have cases involving citizen journalists (governance drive), citizen activists (democratic drive), and citizen producers (co-production drive) (Isin & Ruppert, 2015). In other terms, 'smart citizens' can perform and enact rights through the Internet and digital technologies, but in making rights claims they create openings, contestation, and innovations for existing conventions (ibid.). Reclaiming the 'smart citizen' as a political actor is, therefore, a way of reclaiming citizens' embeddedness into social structures and place.

However, the citizen that emerges from these configurations is a techno-aware one, with a fair amount of cultural and social capital. In this ideological construction, the 'smart citizen' appears to be a subject embedded in and acting through smart technologies, algorithm-led environments, automatic mechanisms of data-value extraction, and digital platforms. As we will debate further in the next chapters, post-political societies frame citizens more frequently in a non-participation modality. 'Non-participation' is underpinned by a strong technocratic impulse (aspects of the city can be treated as technical problems that can be addressed by technical solutions), and notions of stewardship and civic paternalism, where citizens are little if ever consulted in how initiatives are formulated or deployed. Their participation is thus narrowly framed in a very instrumental way (Cardullo & Kitchin, 2019a). Issues inherent in the deployment, adoption, and appropriation of digital participation (through personal social media accounts, electronic voting, civic digital platforms, software coding, etc.) suggest that digital enactments of citizenship can become flawed with a whole set of ethical violations. These all have significant consequences for how citizens are conceived and treated (e.g., as data points; or as subjects to be actively managed and policed; or as consumers), and can work to reproduce and reinforce inequalities.

Moreover, users and cities face unwanted costs in putting 'smartness' into practice. Sometimes, people (or cities) might wish to withdraw, unsuccessfully, and this can create frustrations (or technological lock-ins) – e.g., Google does not allow routing Android phones which become all-Google cages, with less memory available and less choice for alternative privacy-aware software. Another example from everyday life comes from a podcaster and social media influencer who declares, "For years I marked every Facebook ad as spam in order to get rid of them,"[25] leaving one wondering how much digital labour this operation eventually requires and whether it is ultimately worthwhile. Another way of saying this is, there is always an inherent cost and a rather complicated mechanism that might enable and protect privacy in digital societies (Papacharissi, 2010). And as far as withdrawal from smart technologies is possible, this option should be offered as default, or at a push of a button away. In other words, and counterintuitively, the 'smart citizen' appears rather devoid of agency and political capital when acting through digital technologies.

Instead, what will emerge from the discussion on citizenship in modern societies (see Chapters 4 and 5) is the sense in which the journey of citizenship is incomplete – it builds on and expands existing frameworks of civic, political, and symbolic rights claims. Thus, citizenship is a contested terrain which involves political action and awareness, rather than just statements in existing laws or a problem-solving exercise for AI. Ultimately, citizenship is a matter of social justice which belongs to the 'political' in the sense of the performative and participatory stakes and spaces that this concept implies (Cardullo et al., 2019). As D'Ignazio and Klein put it (2020, Chapter 8), "Justice is a journey and the most important part of it is that you stay with the trouble (and hopefully cause some trouble, yourself)."

Concluding remarks

As seen in the previous chapters, within the prevalent neoliberal organisation of the 'smart city' there is a re-orientation of citizenship towards market principles, where the market is meant to act as a "means of regulating and coordinating the activities of numerous actors without direction from a single controlling centre" (Hindess, 2002, p. 140). In the process, neoliberalism shifts citizenship away from inalienable rights and the common good and towards a conception rooted in individual autonomy, freedom of 'choice', and personal responsibilities and obligations, through nudging and normalising behaviour and subtle forms of social control (e.g., Brown, 2016; Ong, 2006; Vanolo, 2016).

In the neoliberal smart city, 'choice' is extended in space and time thanks to the proliferation of interconnected and location-aware devices. The paradox of fostering increased choice with less meaningful participation for citizens is due to the contradictory coming together of forms of technocratic and market-driven governance with poorly understood and practised notions of conviviality, commoning, civic deliberation, resource sharing, trust building, and other face-to-face forms of confrontation and living that make *polis* and communities work. In addition, as the work of city administrations is marketised, deregulated, and privatised, the political and social aspects of citizenship likewise become transformed: instead of right-bearing users of public utility, there are consumers able to select options on the basis of their ability to afford them. Here, the onus to navigate and negotiate the provision of services and levels of access is on the individual, within 'commonsensical' constraints and post-political tropes (sustainability, pollution, planetary urbanisation, congestion, and so on).

While claiming to increase meaningful forms of direct participation, thus, neoliberal governance works within structuring bureaucratic, technological, and ideological path dependencies and representational practices that define a citizenship regime (Cardullo & Kitchin, 2019b). As March (2018) suggests, the 'smart city' discourse on ethics seems to replace the pursuit of social justice with that of the democratization of technology and digital participation. For Hollands (2015) too, the right to use technology has become more urgent than the right to shape the city according to citizens' own needs. So, in order to understand the citizenship regime in operation within the 'smart city' we would need to unpack the "distribution of responsibility between the individual, the community, the market, and the state"; "the rights and obligations, which establish the boundaries of a political community"; and "the governing practices, including modes of citizen engagement and access to the state" (Joss, Cook, & Dayot, 2017, p. 32). All these categories are inherently *political* in that they shake the ideological and technocratic foundations on which the 'smart city' is constructed.

Concurrently, the proliferation of procedural 'models' around ethics of AI and data risks neutralising the impact of rights demands for ethical technology by reducing ethics to compliance and technicalities: checklists, ethics board meetings, tokenistic consultation on secondary issues, PR programs, and marketing stands. 'Ethics washing' practices seem more an attempt by the tech industry to avoid change and evade liability. Therefore, we would need to shift away from an ethics model compliant to neoliberal imperatives of efficiency and privatization which, at best, can tackle issues of customers' satisfaction and value. Ethics requires more than compliance: along with data ethics, we would need data politics (or data justice), history, sociology, and economics of data, good data laws and policy, and relentless activism.

Chapter 4 unpacks these forms of governance at distance from the perspective of the 'citizen'. Thus, a debate on citizenship follows, which frames digital rights claims within a historical and contested tradition of established and emerging rights claims: civic, political, economic, and symbolic 'elements' constituting a modern ideal of citizenship. Drawing on the work done recently with Rob Kitchin and by the 'Programmable City' team of researchers in the last four years, the chapter provides an empirical analysis of the trends of the 'smart city' initiatives in Dublin, Ireland. Like many other cities around the world, the present neoliberal focus of Dublin city (e.g., Coletta et al., 2018; Kitchin et al., 2012; MacLaran & Kelly, 2014) tends to privilege technological solutionism within a framework of marketisation and further privatisation mostly for the benefit of corporate high-tech. As a consequence, citizens' involvement in technologies, initiatives, and plans shifts mostly towards individualised and instrumental positions, such as those of 'consumer' or 'resident' or 'data point': these roles are essentially passive or with very limited informative or consultative power (what we suggest, with Arnstein, are lower forms of 'tokenism' or 'non-participation'). Finally, Chapter 5 concludes this part of the book, critically delving around the instruments meant to deliver a more democratic 'smart city': Living Labs and similar initiatives which are envisioned oft-times as the bottom-up and empowering 'smart city' of the future.

Notes

1 e.g. https://smartdublin.ie/smartstories/smart-bins/
2 e.g. The EIP-SCC has the "bold ambition" to upgrade 10 million lampposts across EU cities with the latest technologies: https://eu-smartcities.eu/initiatives/78/description
3 e.g. www.thejournal.ie/monna-bench-charging-4633425-May2019/
4 e.g. www.chainstoreage.com/technology/facial-recognition-retail-market-research-next-big-thing/
5 For instance, Accenture's 'The Dock' building in the Dublin City's Smart Docklands area contains over 10,000 sensors that "intuitively" control almost everything and generate 1million data points every day: http://smartdocklands.ie/project/accenture-the-dock/
6 The so-called 'Internet of Dildos' is now a booming market of remotely controlled and interconnected devices.
7 Although the Chinese state obviously does also that, as the recent dramatic events in Hong Kong have shown.
8 The "Next Internet" is a mix of centralised cloud computing, Big Data analytics, and Internet of Things.
9 www.scmp.com/tech/china-tech/article/2120913/china-recruits-baidu-alibaba-and-tencent-ai-national-team
10 https://global.ilmanifesto.it/whats-behind-chinas-sharp-eyes/
11 www.nytimes.com/2019/04/14/technology/china-surveillance-artificial-intelligence-racial-profiling.html
12 https://www.theverge.com/2020/6/8/21284683/ibm-no-longer-general-purpose-facial-recognition-analysis-software
13 http://bostonreview.net/print-issues-politics/frank-pasquale-quantifying-love
14 See Kitchin (2019a) The Right to the Smart City. Talk presented at Tilburg University, 14th March. Slides available at: www.slideshare.net/robkitchin/the-right-to-the-smart-city, and more recently: www.rte.ie/brainstorm/2019/0425/1045602-the-ethics-of-smart-cities/
15 See AlgorithmWatch's inventory at https://algorithmwatch.org/en/project/ai-ethics-guidelines-global-inventory and also Alan Winfield's blog: http://alanwinfield.blogspot.com/2019/04/an-updated-round-up-of-ethical.html
16 https://ec.europa.eu/digital-single-market/en/high-level-expert-group-artificial-intelligence
17 Thomas Metzinger is Professor of Theoretical Philosophy at the University of Mainz and was a member of the commission's expert group: www.tagesspiegel.de/politik/eu-guidelines-ethics-washing-made-in-europe/24195496.html
18 www.ilga.gov/legislation/BillStatus.asp?DocNum=1719&GAID=15&DocTypeID=SB&SessionID=108&GA=101
19 e.g., www.vox.com/future-perfect/2019/4/3/18292526/google-ai-ethics-board-letter-acquisti-kay-coles-james; see also this recent case involving Google and its data ethics

board, in the healthcare sector: www.wsj.com/articles/google-quietly-disbanded-another-ai-review-board-following-disagreements-11555250401

20 https://tada.city/en/nieuws/tada-in-practice-jim-stolze-and-aigency/

21 Only last year, for instance, ING Bank was subject to criminal investigations by Dutch authorities regarding the on-boarding of clients, money laundering, and corrupt practices. See: www.cnbc.com/2017/03/22/ing-faces-money-laundering-and-corruption-probe.html

22 This is produced since 2015 by the independent and not-for-profit Ranking Digital Rights (RDR): https://rankingdigitalrights.org/index2019/report/about-index/

23 Notably, Amazon and Alibaba are not included in the ranking as yet

24 https://rankingdigitalrights.org/index2019/report/executive-summary/

25 https://twitter.com/HelenZaltzman/status/1125833987031584768?s=20

4   Citizenship and citizens

Introduction

In the last few years many cities have created and implemented policies and programmes intended to transform them into a 'smart city'. To that end, city administrations, often partnering with companies, have adopted a variety of innovative networked technologies to mediate the management of city services and regulate city life (e.g., city operating systems, urban control rooms, coordinated emergency management response systems, intelligent transport systems, smart grids, smart lighting, sensor-networks, and so forth). These have been complemented by a number of initiatives and services produced and delivered by companies and civic organizations, such as mobile/locative media and mapping and forms of sharing economy (e.g., digital platforms which connect distributed groups of people towards a more efficient use of goods, skills, and services).

Under the rhetoric, funding regimes, and implementation of the 'smart city', there has been a further shift in the neoliberalisation process of city life. This involves steering and rethinking service provision and urban governance towards marketisation, through the five dimensions discussed in the previous chapters: technical solutionism; nudging behaviour; financialisation of service provision; scaling and replication; and creation of smart enclaves. At the same time, apparently 'neutral' technological solutions are activated at the level of the individual, invading, monitoring, and conditioning the flow of everyday life. As seen, both scales of implementation (urban and individual) are often woven into the rhetoric of 'citizen-focused' initiatives. For example, the European Commission has branded its funding programmes for creating smart cities as the 'European Innovation Partnership for Smart Cities and Communities' (EIP-SCC) with a dedicated 'citizen-focus' cluster. Likewise, cities have branded their 'smart city' programmes and initiatives as 'citizen-centric' or 'citizen engaged'. Although projects are implemented in specific contexts, by paying attention to relations, settings, agents, and institutions at a broader scale we can unmask the ideological work that conditions cities in delivering smart city projects for (or rather, on behalf of) 'the citizen'.

Drawing on critical scholarship literature and the research I have done in the last years, the following two chapters unpack the meanings of citizenship in the 'smart city'. The chapters start asking: What kind of citizen has been imagined, operationalised, and activated within new sociotechnological assemblages that form the 'smart city'? Is the advent of the 'smart citizen' – a subject embedded in and acting through smart technologies, algorithm-led environments, automatic mechanisms of data-value extraction, and civic digital platforms – perhaps changing the notion of citizenship inherited from liberal, social, and symbolic rights claims tradition? Are new digital rights claims (Isin & Ruppert, 2015) needed to reassert established rights and reshape citizenship in new political configurations – and are these enough? This chapter sets out citizenship ideals, in order to put the 'smart citizen' in the context of rights claims as an historical and conflictual – thus, inherently political – process. It offers an overview of citizenship as an evolving concept which shapes and is shaped by the political organisation of society. Further, the chapter presents a composite case study of 'smart citizens' in Dublin, Ireland, using the heuristics Rob Kitchin and I (Cardullo & Kitchin, 2019a) have built in relation to the 'actually existing' citizen in the smart city. Although contextual to the city under examination, which is always a complex amalgam of many different institutions and governance practices, our 'scaffold of smart citizen participation' is a flexible and detailed enough heuristic to deliver a working toolkit for policymakers, activists, and smart city scholars, delineating the variety of roles citizens perform as they become, willingly or not, part of complex and evolving technological assemblages. Citizens in the 'smart city' can be steered, nudged, and controlled as passive data-points; they can browse, consume, and act as consumers. Or they can be residents in smart enclaves or workers and players for digital hubs. Whatever role(s) they take, we argue that citizens' civic engagement in the 'smart city' is rather limited, and often in the form of a participant, tester, or player who provides feedback or suggestions; rather than taking on more active and deliberative roles such as a proposer, co-creator, decision maker or leader who has stakes in the technology or power to steer its aims or adoption.

There is an evident disjuncture between rhetorical claims around 'citizen-centric' smart cities – combined with an ideological belief that technology is the optimal solution for everything – and the actually existing roles citizens perform in the neoliberal 'smart city'. This gap between aims and adoption creates a paradox for the neoliberal subject. The onus to navigate and negotiate the provision of services and levels of access is, in fact, placed on the individual, based on their personal, social, political, and economic capital. At the same time, neoliberal smart urbanism frames citizen's roles within commonsensical constraints (for instance, with regard to the reduction of externalities brought by urbanisation: pollution, congestion, depletion of natural resources, climate change, etc.). According to Han (2017), the neoliberal subject is not a 'labourer' any more, but a 'project' of biopolitics. In order to unpack how this 'project' is enabled and works within the 'smart city', we first need to trace the journey of citizenship in contemporary capitalism: in the current neoliberal framework through which the 'smart city' ideal is implemented, real-time algorithm-led decisions are generally preferred over political discussion and agonistic processes of governance.

The journey of citizenship

In discussing citizenship, I ground the concept in the everyday practice of urban living and the materiality of the 'right to inhabitation'. In my and critical scholarship's view (Purcell, 2003), citizenship ought to be considered much more than a status linked to birth and nationality. Rather, citizenship substantiates in a set of relationships within places of inhabitation, between other fellow citizens, and with regards to city institutions and urban governance of infrastructures. Ultimately, the very sites and practices through which citizenship is performed are said to give the actually existing citizen its meanings (Staeheli, 2011). I will return to this ideal of citizenship in the second part of the book, when debating a more normative framework leading to a 'smart city' ideal alternative to the current dominant version. Before that, it is useful to trace the development of the concept as modern societies change along organisational, legal, and sociopolitical processes. This is a wide area of scholarship which, intuitively, covers different fields of research, in laws and community studies, but also in urban geography, critical political economy, politics, and cultural studies. One way of doing this is to map the historical evolution of citizenship as a concept: in the remainder of this chapter, I will debate this briefly with particular reference to Western societies, although also mentioning the variance through which these ideals have been translated and reworked by postcolonial scholars. In his classic text, *Citizenship and Social Class*, T. H. Marshall (1992 [1950], p. 8) suggests considering three sets of rights that define the citizenship status of citizens as "denoted by history even more clearly than by logic." These are "parts or elements" that constitute the modern concept of citizenship, that is, they are not complementary or optional features of citizenship but rather co-constitutive. Consequently, citizenship is composed by a civil/legal element (e.g., the right to own property; freedom of speech; liberty of the person, and the right to justice), which concerns mostly the juridical person and the courts of law; by a political element (e.g., the right to vote and participate in the exercise of political power and in related assembling), which concerns citizens' active participation in the political life of city or nation as both voters and persons invested with a certain authority; and by a social element (e.g., the right to a certain level of economic welfare and security "and to live the life of a civilised being according to the standards prevailing in the society" [ibid.]), which becomes evident in the educational and welfare systems. The point of Marshall's exposition is that these elements of citizenship have been separated from each other and moved up on a more-than-local scale (national and, more recently, international), such as parliament, welfare state, and high courts. Moreover, the separation of citizenship into 'elements' becomes an almost chronological affair, corresponding to the formation of the liberal law and the affirmation of Westphalian nation states in the 19th century, all the way through the political and social rights movements of the last century.

To this set of rights have been added more recently cultural/symbolic rights. These concern recognition, respect, and protections with regards to identity (gender, race, sexuality, disability, age, and faith). Similarly, liberal theorist Waldron (discussed in Attoh, 2011) divides, historically and conceptually, rights claims and demands into three generations: liberal tradition (freedom, privacy, speech); socioeconomic tradition (housing, welfare, income); and so-called 'solidarity rights' (minorities, language, environmental values, culture). The first two categories immediately present the critical issue of their interdependence, which is a matter of much contestation among different scholarly traditions (the Liberal and Marxist ones, for instance). Indeed, Marshall's project was an effort to trace the history of citizenship as an all-around concept which counterweighted somehow the inequality of social classes, eventually "altering the patterns of social inequality" (1992, p. 44). His optimism with regards to citizenship finds room in his aim of challenging the patterns of capitalist accumulation: "economic inequality has been made more difficult by the enrichment of the universal status of citizenship" (p. 45): thus, Marshall frames these elements of citizenship as a national project. Instead, the 'right to the city' movement aims at establishing the urban, or the municipality, as the main locus of inhabitation – for Marshall, this happened before in some medieval towns, which were able to implement "genuine and equal citizenship" (1992 [1950], p. 9). As I will probe the concept further in the second part of the book, I will ascertain the right to inhabitation as central to the 'right to the city'. "Democracy begins where you live," has been the motto used by the Municipalist Confluences in Spain (Rubio-Pueyo, 2017). In this way, I want to convey the importance of local administration and citizens' direct experience in the expansion and transformation of institutions towards a more democratic and just society. In other words, rights claims are always contextual and embedded in the localities and through the determinants of space-based analysis.

The above periodisation in the development of rights claims presents overlapping stages, and some obvious disjunctures: notably, women's political rights were gained only much later than men's, or civic rights of Blacks and minority groups have been asserted only in regards to White people's claims. These frictions suggest immediately that rights claims have been contested and, actually, emerge as an ongoing struggle throughout the entire history of modern societies: in this, they are inherently political. For Mouffe (1999), it is exactly this contested terrain of rights claims that renders democracy functioning and dynamic. Democracy ought to strive towards forms of active participation which she calls "agonistic pluralism." This is different from forms of deliberative democracy (of the majority, of the 'public sphere', in Habermas's sense) and from liberal tradition (of individual rights claims, of the voter). Agonism, in fact, forges a notion of citizenship which is *relational* to their community of interests and *plural* with regards to participatory deliberations. In her own words, democratic theory needs to acknowledge "the ineradicability of antagonism" and thus the impossibility of achieving a fully inclusive rational consensus, proper of the deliberative tradition (Mouffe, 1999, p. 756).

From this historical and layered perspective, citizenship appears as "a set of practices (cultural, symbolic and economic) and a bundle of rights and duties (civil, political and social) that define an individual's membership in a polity (usually a nation-state)" (Isin & Wood, 1999, p. 4). But, what does this mean when most people's livelihood, their everyday practices of living the city, especially in the Global North, might depend on access to information, communication, and coding (Lash, 2002)? For instance, now that the Internet of People and Internet of Things have merged into what we might call the 'smart city', access to a reliable and affordable Internet is not just a matter of geographical inequalities, between the city and the rural or between areas of the same city, and probably neither solely an issue of rights claims. It is instead a matter of basic participation in civic life, one gateway for the 'right to the smart city': whether we like it or not, digital services are becoming the privileged or only way to pay taxes or parking tickets, to apply for a job or complete school homework, to register a birth or vote: in other words, the way in which people, especially in the Global North, inhabit cities every day. In this regard, Isin and Ruppert (2015) maintain that digital acts and rights claims through, or about, digital technologies – such as witnessing, hacking, and commoning – are not simply 'virtual' acts, somewhere in the cyber-domain. They are instead very real, enacted through social skills and cultural capital. They are enactments which carry a 'performative force' that endures. As a consequence, digital acts and digital rights claims can produce forms of affect: frustration, joy, or anger can be some popular emotional investments likely to be linked to participation to, or exclusion from, digital infrastructures (Larkin, 2013; Rodgers & O'Neill, 2012).

In my view, however, 'digital rights' are not an additional set of rights to be added to other 'elements' of citizenship, but an overlapping layer which reinforces Marshall's exposition of rights. Established rights claims are *also* 'digital', or having *increasingly* something to do with the 'digital', such as: political rights expressed, inhibited, or enhanced through digital platforms or electronic voting; civic rights shaped and regulated deeply through algorithmic processing, such as apps denoting mortgage accessibility, crypto-ledgers adjudicating property deeds, or financial transactions happening in milliseconds; or symbolic rights determined by bias and abuse through an AI-led decision-making process. For instance, the digital right claim to privacy is much debated around ethics of data, artificial intelligence, and algorithms. Privacy is an established liberal right claim both in the legal realm and in everyday practices (e.g., by shutting doors and windows, lowering voices, or hiding from unwanted gazes); it is so pervasive that it was included in the UN Universal Declaration of Human Rights, 1948: art. 12 (Diggelmann & Cleis, 2014). But privacy is also a "soft term," according to Jaap-Henk Hoepman (2019) who suggests, in his manual of privacy-by-design strategies, "[privacy] is partly about ethics and partly about legal aspects. That makes it harder to implement than more tangible software attributes like security or performance." We see here how 'the digital' only juxtaposes with other dimensions of rights claims rather than representing a new set of claims.

Thus, I would suggest splitting 'digital rights' into two broad dimensions: the ontological and the procedural. We can think of the first dimension as the set of civic, political, social and symbolic rights which is expressed through, or gets enabled by, digital devices and data. In this case, the digital rights associated with it are the same constituting citizenship rights in other fields, such as privacy, security, information self-determination and neutrality, just to mention some. To this, I would add 'commoning', when digital resources and data are shared and claimed as a collective wealth: however, commoning in this case is a modality *of* the digital or *through* the digital (e.g., by way of using free and open source software) rather than a right claim *per se*.[1] Instead, the *procedural* dimension of digital rights claims looks at enabling active forms of citizens' direct participation in the making and governance of cities, through and because of digital devices and data. To this dimension would belong forms of direct feedback and consultation to city deliberation or budgeting via, for instance, digital civic platforms (where these are enabled) or open data sets (when these are useful).

Smart citizens?

The fostering of the *procedural* dimension of digital rights resembles, often times, an exercise of digital democracy which aims to enhance citizens' information, consultation, and feedback. These informed and digitally active citizens are consequently framed as 'smart citizens', capable of interacting with institutional *systems*, via informational *systems*, with the aim of influencing decisions taken by the various managerial or governmentality *systems*. The rhetoric used in the previous sentence highlights exactly how the decision-making process is intertwined with, and subject to, the many other variables that establish these various *systems*, which might enable or hinder citizens' participation. Notions of governance and engagement mix with corporate and institutional goals, which make the decision process fairly difficult to become open, democratic, and amenable. As a consequence, the final outcome for the 'smart citizen' is that meaningful participation leading to effective change is unlikely to happen. Or at least, this is negotiated through contingent institutional processes which are not automatic or autonomous. In other words, I believe that there is nothing deterministic about, for instance, open access data sets or open source software in making a decisional process more or less democratic.

To this end, the next section maps and evaluates actually existing citizens' roles and their effective participation in the development of the 'smart city' with a multiple and heuristic case study of Dublin, Ireland. In a recent paper, Kitchin and myself (Cardullo & Kitchin, 2019a) reworked Sherry Arnstein's (1969) well-known 'ladder of citizen participation' to examine the various citizen roles enacted across 'smart city' initiatives. What we found was that citizens most often occupy non-participatory, consumer-centred, or tokenistic positions; as a consequence, they are framed within political discourses of stewardship, technocracy, paternalism, and the market, rather than being active and engaged participants. If there is civic engagement, this is in the form of a tester or player who provides feedback or suggestions, rather than being a proposer or a leader. And yet, most 'smart city' initiatives claim to be 'citizen-focused' or 'citizen-centric'. In my view, the disconnection between discursive intent and reality is caused by two factors.

First, initiatives that were critiqued for their top-down and technocratic nature have sought to silence detractors by re-branding their endeavours as 'citizen-centric', while keeping the central mission of capital accumulation and governance intact (Kitchin, 2015). The process of marketisation of public services, digital infrastructures and smart technologies points exactly to the direction of harnessing forms of value other than the public value and the common good. At its best, a sort of 'ethics washing' (e.g., Wagner, 2018) has been enabled via various ethics audits and committees, or via the growing professional profile of the ethicist in residence – a sort of *deus-ex-machina* who will guarantee the safeguarding of citizens' rights, primarily individual consumers' value or customer's satisfaction, while also guaranteeing compliance with market priorities of efficiency.

Second, funding programmes designed to encourage city administrations to deliver the 'smart city', such as the European Commission's EIP-SCC, structurally preclude any serious intent to engage citizens in the formulation of projects (Cardullo & Kitchin, 2019b). Putting together a large, multimillion euro bid is a time-consuming, complex, and largely unfunded task; and adding 'non-expert' citizens into the process creates significant additional overheads. What this means is that in most cases the focus, objectives, and solutions are set *before* any problems and suggestions from citizens can be taken into account, and it is only when the funding is in hand that engagement occurs with local communities. Such citizen engagement has to meet predetermined milestones and fulfil the deliverables of the contract, meaning participants have limited scope to subsequently reframe the initiative around their concerns and desires. In such formulations, citizenship operates largely as an empty signifier, often calling for "citizen inclusion" or searching for the "missing citizen," with the underlying neoliberal ethos and mode of governmentality remaining unchanged (see Hill, 2013; Sartori, 2015; Shelton & Lodato, 2019).

Despite the drive to create 'citizen-centric' smart cities, then, to date there has been little critical conceptual scrutiny as to how citizens are imagined and engaged by different 'smart city' technologies and the model of citizenship enacted within 'smart cities' – although there are case studies and theoretical scrutiny of citizen engagement in crowdsourcing (e.g., Gabrys, Pritchard, & Barratt, 2016), in participatory planning (de Lange & de Waal, 2016), in the device-enabled shift from user to consumer (Fuller, 2017), or more broadly in terms of 'smart city' policy and strategies (Cowley, Joss, & Dayot, 2017; Datta, 2018; Ribera-Fumaz, 2019; Shelton & Lodato, 2019; Vanolo, 2016).

Cowley et al. (2017) identify four modalities of 'publicness', which denote how citizens are positioned within the 'smart city': 'service user', in which citizens are framed as the consumers of services; 'entrepreneurial', in which citizens are actively enrolled into co-creating and innovating; 'political', in which citizens take an active role in decision-making and deliberation; and 'civic', in which citizens take part in grassroots community activities that are not directly oriented towards market activity. They note that there is a significant variation of publicness across initiatives and cities, mostly favouring 'service users'.

In contrast, Datta (2018) suggests the ambiguous role of concepts like 'smart citizenship' for the postcolonial subject who, in her view, continually operates in the grey zones of informality and everyday re-appropriation of technologies. Context, as always, reigns in this configuration since Indian citizenship swings between a uniform civil society of universalised rights and "the demand for differentiated rights for those who have historically faced social and economic marginalisation in society" (p. 4). As such, Datta (2018) sees smart citizenship as an amorphous and dialectic identity across three overlapping vectors: 'enumeration' through citizen consultations of a digital population, 'articulation' of how to become smart citizens, and 'breaches' to particular forms of authority and power across digital and urban publics. The 'smart citizen' appears less a connected subject (a euphemism used for an elite citizenship built on class, religious, and caste privilege) than a 'chatur citizen' – a postcolonial subject who shares the same space but becomes also an embodiment of situated and discursive politics across digital and subaltern publics (ibid.).

Further, Shelton and Lodato (2019) note in their study of Atlanta 'smart city' programme that, while the city administration and companies attending their events often talked of producing a 'citizen-focused smart city', in practice citizens were included as two empty signifiers. The first is a 'general citizen'; a catchall term for a community of seemingly homogeneous recipients or consumers of services. Here, the 'smart city' operates within the framework of stewardship (delivering on behalf of citizens) and civic paternalism (deciding what's best for citizens), rather than citizens being meaningfully involved in the vision and development of the 'smart city'. The second is the 'absent citizen', referring to all those diverse communities that hold identities, values, concerns, and experiences differing from the 'general citizen' (who is largely framed as white, male, heterosexual, ablebodied and middle class). To this can be added a third figure, that appears often in other 'smart city' documents and programmes, the 'active citizen'; an entrepreneurial citizen that builds civic tech for community development through hackathons and other events (Hemment & Townsend, 2013).

The 'active citizen' is a particularly important and growing character in the 'smart city' landscape of today. For instance, in order to increase citizens' participation and involvement in 'solving' local issues, the Living Labs concept has been introduced as a bottom-up approach to the smart city. Thus, Living Labs are intended to operate as multi-stakeholder endeavours that sometimes include local residents, acting as a counter-weight to the techno-centric, top-down approach to the 'smart city' initially forwarded by big business. Ultimately, the ambition for some is that the 'smart city' will eventually boast a model of governance in which "a community assumes political and expert management over its infrastructures" (Corsin Jimenez, 2014, p. 343). Being citizen-led or citizen-engaged in the 'smart city' does not necessarily confer notions of citizenship or rights to the digital city, or produce a new digital urban commons. The development and use of participative 'smart city' software interfaces and Living Labs seem, in fact, to produce an 'ideal citizen' that willingly subscribes to the idea(l)s of technological solutionism promoted by the 'smart city' discourses and acquires the cultural capital necessary to communicate or tinker with smart technologies (Cardullo et al., 2018). While often worthy and much more preferable to top-down forms of governance, it is unlikely that Living Labs can fulfil 'citizen-centric' smart city goals, without these being explicitly rooted in notions of citizenship and community ownership, rather than citizen participation and civic paternalism. The root to this, we argued (ibid.), is within a communal city model of citizen engagement which I discuss more in detail in the second part of this book.

To this increasingly popular initiative at the heart of the 'smart city', Living Labs, I dedicate the next chapter which concludes the first part of the book. The next section presents a detailed discussion of 'smart citizens' in the context of Dublin, Ireland, where an intensive empirical exploration of the various 'smart city' initiatives has been undertaken by the 'Programmable City' team at Maynooth University Social Science Institute (MUSSI). To this project, I devoted two years of my professional life working with other researchers and the Principal Investigator of the project, Professor Rob Kitchin.

Actually existing smart citizens

While the previous sections set out 'citizenship' in its historical context, this section looks more in detail at how forms of governance at a distance – indeed automated through various forms of countering, measuring, and monitoring – have been promoting a model of participation that is rooted in pragmatic, instrumental, and paternalistic discourses and practices in which the 'citizen' takes (even simultaneously) many roles. Here, I present the heuristics around citizen's participation in the 'smart city' with regards to Dublin, Ireland, as conceived with Rob Kitchin (Cardullo & Kitchin, 2019a). I want to use this to work through common applications and adoption of smart technology in cities. Notwithstanding the contextual and local differences in the 'smart city' implementation (not two projects or cities or even stages of the same project are the same!), there are commonalities and challenges offered by the rolling out of algorithm-enabled and often networked technologies which we only recently started to understand and unpack in detail (Greenfield, 2017; O'Neil, 2017; Zuboff, 2019). One of the difficulties of this exercise, of course, is the relative opacity of such technological assemblages and digital platforms. These, while becoming increasingly ubiquitous and invasive and demanding individual participation and engagement by default for various forms of data extraction and monitoring, remain substantially privately owned and protected by patents or copyrights, black-boxed and away from public political oversight (see Fuller, 2017).

The 'scaffold of smart citizen participation' is a heuristic tool to compare and evaluate different projects from the perspective of citizen's roles (see Table 4.1). Beyond the powerful rhetoric of the 'smart city' discourse, the scaffold is a map of inclusion and participation in 'smart city' projects through which scholars and stakeholders can better understand how the 'citizen' is involved, and in what capacity, in any existing and forthcoming 'smart city' initiatives. Rather than being exhaustive, therefore, the scaffold wants to provide the basis for the formulation of new avenues of enquiry – for example, with regards to comparative analysis of different institutional arrangements and scales in the delivery of smart

Table 4.1 Scaffold of smart citizen participation

Form and Level of Participation		Role	Citizen Involvement	Political discourse/ framing	Modality	Dublin Examples
Citizen Power	Citizen Control	Leader, Member	Ideas, Vision, Leadership Ownership, Create	Rights, Social/ Political Citizenship, Commons	Inclusive, Bottom-up, Collective, Autonomy, Experimental	Code for Ireland, Tog
	Delegated Power	Decision maker, Maker				Civic Hacking, Hackathons, Living Labs, Dublin Beta
	Partnership	Co-creator	Negotiate, Produce	Participation, Co-creation		Fix-Your-Street, Smart Dublin Advisory Network
Tokenism	Placation	Proposer	Suggest			CIVIQ, Smart Stadium
	Consultation	Participant, Tester, Player	Feedback	Civic Engagement		Dublinked, Dublin Dashboard, RTPI
	Information	Recipient			Top-down, Civic Paternalism, Stewardship, Bound-to-succeed	Smart building/ Smart district
Consumerism	Choice	Resident, Consumer	Browse, Consume, Act	Capitalism, Market, Neoliberalism		Smart meters, Mobile/locative media
Non-Participation	Therapy	Patient, Learner, User, Product, Data-point	Steered, Nudged, Controlled	Stewardship, Technocracy, Paternalism		Dublin Bikes, Smart Dublin
	Manipulation					Traffic control

city projects; or to the timeline through which projects are prepared, funded, and institutionalized; or to the actual existing spaces for feedback and adjustments within such projects. As way of illustration, we used the scaffold to assess the 'citizen-centric' nature of 'smart city' initiatives in the city of Dublin, Ireland.

With the exception of some 'citizen power' initiatives, all levels of the scaffold are consistent with neoliberal citizenship ideals and its emphasis on personal autonomy and consumer choice. Individuals perform certain roles and take some responsibility for their own life chances (entrepreneurial self), in the context of increasing marketisation and privatisation of services and infrastructures ('retreat' of the state, or rather the reformulation of its functions and responsibilities according to market imperatives, in a climate of austerity policy). While citizen participation is potentially diverse, it is most often framed in a post-political way in which the citizen's capability of providing feedback, negotiation, and creation remains instrumental rather than normative; that is, unable to foster a new political vision for the future smart city. In other words, citizens are encouraged to help provide solutions to practical issues – such as producing an app, or providing feedback on a development plan, or performing certain roles/responsibilities – but not to challenge or replace the fundamental political rationalities shaping an issue or plan.

Instead, most citizens are empowered in the 'smart city' by technologies that treat them as consumers or testers, or people to be steered, controlled, and nudged to act in certain ways, or as sources of data which can be turned into products. In other words, 'smart citizens' perform within the bounds of expected and acceptable behaviour, rather than transgressing or resisting social and political norms. Individual responsibilities are understood within the framework of unlimited consumer choice, an important component of the neoliberal approach to behavioural change that views regulation and restriction of choice with suspicion (Clarke, Newman, Smith, Vidler, & Westmarland, 2007). Here, 'choice' lies within the narrow modes of passive citizenship, compliance, and the 'citizen-consumer' framework. The latter hints at "a populist discourse that articulates the 'consumer interest' against the power and interest of the 'producers'" (Clarke et al., 2007, p. 143) (thus, the rhetoric on disruptive technologies such as Airbnb or Uber). As such, 'citizen involvement' expresses a form of neoliberal citizenship not grounded in civil, social, and political rights, or in the promotion of public or common good, but rather in individual autonomy and 'choice'.

Claims concerning the production of 'citizen-centric smart cities' thus appear to be largely tokenistic, with city administrations and corporations still owning and controlling urban governance and services, and 'smart city' initiatives being used to enact a form of technologically-led entrepreneurial urbanism (Hollands, 2008; Kitchin, 2015; Swyngedouw, 2016). The chapter aims to advance the notion of fragmented citizenship in advanced capitalism (Isin, 2000; Ong, 2006), adapting it to the existing and yet to come 'smart cities'. It highlights how policy becomes instrumental to the neoliberal ideal of 'citizen-centric smart city', which ultimately fosters the slipping away of the citizen as a political subject that holds a set of rights and entitlements (although at different times and in different spaces), to much weaker socioeconomic and legal positions.

The scaffold of 'smart citizen' participation

In 1969, Sherry Arnstein published a highly influential paper on the ways in which citizens are involved in the planning process and regeneration programmes. Her thesis was that planning is a top-down, technocratic exercise that takes little account of citizens' views or desires. Arnstein formulated a conceptual ladder with eight rungs "corresponding to the extent of citizens' power in determining the end product" (p. 217). On the lower rungs we find forms of 'non-participation' ('manipulation' and 'therapy'), which are designed to direct and educate people in a top-down, formal manner, steering and controlling them. She then defines 'tokenism' ('informing', 'consultation', and 'placation') as a form of participation in which people have a voice and some degree of autonomy, though they are rarely able to change directly the *status quo* of decisions and plans already taken elsewhere. The final three rungs concern 'citizen power': 'partnership', in which citizens can take an active participative role and share decision making with dominant power-holders; 'delegated power', in which citizens are full actors and have a dominant decision-making role; and, 'citizen control', where "have-not citizens obtain full managerial power" (1969, p. 217).

Thus, in Arnstein's formulation, the quality and depth of citizen participation in planning is rooted in access to power. Although she never defines power, Arnstein maintains the control of power has significant implications for the socioeconomic advancement of the "have-nots" and thus embodies the potential to transform "nobodies" into "somebodies" (p. 217). Participation is linked to power to the extent that it can induce "significant social reform," affecting the outcome of a process and eventually redistributing "the benefits of affluent society," rather than being only an "empty ritual" (p. 216). In other words, for Arnstein, participation through power can work by reflecting an ideal of society that is more equal and just, and enabled through more democratic planning and decision-making practices.

Through critical reflection on how citizens were positioned in each of Dublin's 'smart city' initiatives, we started to match these endeavours to the 'rungs' on Arnstein's ladder – to deductively test both the validity of the theoretical frame *and* the extent to which 'smart cities' are 'citizen-focused' (Cardullo & Kitchin, 2019a). What became apparent was that while the ladder had utility, it also had some limitations. Using our case examples, we then began to iteratively reconstruct the ladder into a scaffold, reflecting on the roles played, the form and nature of citizen involvement, and the underlying political discourse. In this sense, the scaffold was rebuilt inductively through the case study material. In the final stage, we used the scaffold as a heuristic to assess how citizens are conceived and positioned within Dublin's 'smart city' initiatives, the diverse roles they play, and the extent to which initiatives are grounded in and reproduce the discursive and material practices of the neoliberal 'smart city'.

Our field site, Dublin, is a city that promotes itself actively as a 'smart city' through its Smart Dublin[2] office (a unit shared among the four Dublin local authorities to coordinate and promote its 'smart city' mission) and has rolled out a number of mainstream 'smart city' initiatives, as well as acting as a test-bed for many more in development (Coletta et al., 2018). Since the late 1980s, Ireland has embraced the tenets of neoliberalism, creating a political and economic model that blends American neoliberalism (minimal state, privatization of public services, public-private partnerships, developer/speculator-led planning, low corporate and individual taxation, and light regulation) with aspects of European social welfarism (developmental state, social partnership, welfare safety net, high indirect tax, EU directives and obligations) (Breathnach, 2010; Kitchin et al., 2012). In political economy terms, the city is thus similar to many European and American cities which have pursued neoliberal, entrepreneurial, and competitive strategies, including a variety of 'smart city' initiatives. Smart Dublin promotes itself as "Open, Engaged, Connected," where 'engaged' relates to citizen engagement. Although Smart Dublin has been "somewhat successful in creating a smart city narrative and branding Dublin as a smart city," its policy direction reflected "the wider neoliberal ethos of government in Ireland" (Coletta et al., 2018). The focus on more instrumental concerns, such as improving city services and economic growth in a climate of austerity politics and job cuts, has meant that citizens are ultimately considered the beneficiaries of the 'smart city', better informed and maybe receivers of better services but not the political subjects which actively participate, co-create or make decisions.

Our initial reworking of the ladder was to add a ninth rung to the level of participation column: 'choice'. As we detail below with respect to our case study, this is to recognize that in the 50 years since Arnstein was writing, states have embraced neoliberalism, with city services and infrastructures being increasingly marketised (treating citizens as customers) (Clarke et al., 2007) and privatised (corporations owning key city assets and performing many key roles) (Brenner & Theodore, 2002). Although our case study remains invariably limited by localised contingencies of place-based analysis, there are intruding general patterns in the 'smart city' discourse and development which is worth considering. Our proposition is therefore that 'the scaffold' can be used as a heuristic tool by scholars and stakeholders in other cities, too, in order to critically evaluate the citizen-focused nature of smart technologies and projects beyond the rhetoric offered by the 'smart city' discourse.

A prime way in which a citizen interacts with the 'smart city' is as a 'consumer', selecting which services to acquire from the marketplace of providers – or, in the case of free-to-use apps, swapping personal data for. The second role the citizen performs at this level is that of 'resident', especially among those who can afford the purchase/rent price by choosing to live in a 'smart building' or 'smart district', spaces that are often exclusive and gated communities. Consumerism in the 'smart city' is 'citizen-centric' in as far as it seeks to provide a selection of information and services from a range of entities that fulfil a need. We have therefore slotted it into our scaffold between 'Non-Participation' and 'Tokenism' (Cardullo & Kitchin, 2019a).

Our main alterations have been to add a number of related columns, spanning Arnstein's rungs – hence, our use of 'scaffold' rather than 'ladder'. The first column added relates to the role expected of/adopted by citizens with respect to 'smart city' initiatives: by systematically analysing a series of cases in Dublin and elsewhere, we have identified 16 'citizen' roles that shift from passive and lacking control to active and responsible. The second column added concerns the form of citizen involvement and the nature of their engagement, varying from forms of coercion through to visioning and steering initiatives. The third additional column refers to the political discourse used to justify and drive the various forms, levels, roles, and involvement of citizens.

The final additional column is the modality in broad terms as to how citizens are positioned vis-à-vis the 'smart city'. In the lower half of the scaffold, initiatives are most often *top-down* in conception, being devised by city administrations or corporations, and are broadly underpinned by notions of stewardship and civic paternalism (see Clark & Shelton, 2016). These projects are 'bound-to-succeed' in the sense that there is an expectation that these initiatives will deliver on their promise to produce a smarter city, and not waste taxpayers' money or shareholder investment. In contrast, in the top half of the table, initiatives are more *bottom-up* in conception, being devised in part or in whole by various citizens or groups, and are more collective in how they operate. These initiatives are more experimental in nature and it is understood that they might fail to create a long-term, sustainable outcome.

Non-participation

'Non-participation' occurs when citizens are nudged and steered towards specific sets of behaviour, practice, and conduct. This can be the case for interventions that require very little input from citizens other than to use or experience an algorithmically-mediated service for the purposes of governmentality; for example, the production of big data in order to intensify "the extent and frequency of monitoring and shift the governmental logic from surveillance and discipline to capture and control" (Kitchin, Coletta, & McArdle, 2017, p. 3). Here, citizens become subject to a modulation of their actions through software-mediated systems designed to produce particular regulatory outcomes that actively shape behaviour.

For example, in the case of Dublin, traffic flow is regulated by the Traffic Management and Incident Center (TMIC) and its use of SCATS (Sydney Coordinated Adaptive Traffic System) (B. McCann, 2014). SCATS is an automated and adaptive system whose primary role is to manage the dynamic timing of signal cycles and phases at road junctions in order to ensure the optimal flow. The system automatically calibrates the cycles and phases dependent on a set of programmed rules and the flow, speed, and density of traffic for each lane of traffic in previous cycles and phases (as measured in real time by a network of 800 inductive loop sensors) (Coletta & Kitchin, 2016; B. McCann, 2014). In addition, the TMIC has access to

380 CCTV cameras, a small number of traffic cameras, a mobile network of circa 1,000 bus transponders, phone calls and messages by the public to radio stations and operators, and social media posts (Coletta & Kitchin, 2016). Citizens and their vehicles become 'data-points' in a fluctuating system, with the data generated used to calibrate the system and traffic flow. Information from the system is also pushed out to citizens via apps, the Dublin Dashboard,[3] real-time passenger information at bus stops, and on-street signs stating numbers of vacant spaces in car parks, which nudge decision making with respect to choice of route and parking. In addition, these data, along with those generated from other sources (such as locative media) can be mined for insights, traded with and between data brokers, and conjoined with other data for the purposes of social sorting, predictive profiling, micro-marketing, anticipatory governance, and city planning (Kitchin, 2014a). In other words, citizens using algorithmically-mediated services can become 'data products', raising a series of ethical questions concerning over-extended and intrusive surveillance as well as privacy and predictive privacy harms (Kitchin, 2016b).

Non-participation in the 'smart city' is underpinned by a strong technocratic impulse (aspects of the city can be treated as technical problems that can be addressed by technical solutions), and notions of stewardship and civic paternalism, where citizens are rarely if ever consulted on how initiatives are formulated or deployed. Within paternalism, the state, commercial organisations, and local authorities can prescribe the behaviours that individuals need to adopt to make this a reality. Behavioural change becomes more than selling a message, rather an engaged conversation about the future of urban sustainability. Here, citizens' participation is narrowly framed in a very instrumental way.

Consumerism

According to Fuller (2017), the shift from a *user* of technologies to *consumer* is aided by the transformation in the design and functioning of computational devices, from personal computers to cloud- and platform-based economies on the Internet. There are now thousands of app-driven services designed to transform city living. The vast majority of these are owned and operated by private corporations who utilise digital technologies to deliver new services, based on a combination of location, real-timing, identity, and algorithmic profiling. These services are often disruptive, radically altering established markets. For example, a certain version of the sharing economy is transforming the taxi (e.g., MyTaxi) and accommodation (e.g., Airbnb) industries, as well as employment practices (e.g., the gig economy).

In addition, people can embrace a 'smart' lifestyle by becoming a resident in a smart building or district. Such buildings and areas are often served by multiple 'smart city' technologies designed to enhance the lives of residents through improved security, energy and waste services, and transportation and parking options. In Dublin, the 'Silicon Docks' area of the city – a special development zone being regenerated through a mix of high-end offices and residential apartments – has been designated a 'smart district' (Heaphy & Pétercsák, 2016). Home to the European headquarters of companies such as Google, Facebook, and LinkedIn, the area has become a testbed for new smart technologies and acts as a means to attract additional inward investment (especially from Internet of Things companies and 5G enthusiasts). Much of the space created is privately owned and managed rather than being public space, with such developments operating for the benefit of their owners and counter to that of an urban commons. A 'smart citizen' in such developments is a high-income consumer probably seeking an exclusive property investment with the latest technological trimmings (see also Mattern, 2016; Shin et al., 2015).

In general, services delivered through smart technologies are designed and operated with limited involvement by citizens other than as users. If citizens are involved, it is usually as feedback during requirement's analysis in the design phase or as beta-testers of products in the production phase. Here, feedback is used to tweak already conceived designs, rather than to form the bedrock for design thinking. Like 'non-participation', 'consumerism' is then undergirded by a strong technocratic framing. It also has strong notions of stewardship and paternalism, with the market largely determining what is in the best interests of citizens.

Tokenism

'Tokenism' concerns various degrees of public engagement and citizen voice. In its lower form, it consists of telling citizens where they can access open data that, on the one hand, inform them as to what is happening in the city, and on the other, can be repurposed to form the input for citizen-created apps. In Dublin, Dublinked[4] – an initiative co-owned by the four local authorities – is the city's open data store, sharing a mix of administrative and operational data, including some real-time datasets related to transport and environment. Much of these data, along with statistical data and administrative data published by other government agencies, are made available to the public through the Dublin Dashboard as interactive maps, graphs, and apps. Such information can be used to shape decision-making and can also be used to create transparency and accountability with regards to the actions and decisions of administrations (a key argument of the open data movement). However, while 'informing' can be highly useful, it is often uni-directional, with limited or no channels for providing feedback. Moreover, information is often available after key planning and decision-making processes have occurred, leaving little or no room for change.

In its higher forms, tokenism constitutes 'consultation' and 'placation'. In 'consultation' citizens are requested to provide feedback representing their views through various forms of social media and online tools for citizen consultation (de Waal, 2014; Seltzer & Mahmoudi, 2013). Indeed, even citizen engagement can become a "lucrative and expanding business," as declared the CEO of a city platform app service company I interviewed during the Barcelona annual Expo on smart cities (Smart City Expo World Congress, 2017) (see Cardullo & Kitchin,

2019b). In the Dublin case, an example would be the use of CIVIQ, an online consultation tool that enables citizens to comment on and discuss draft county development plans.[5] In 'placation' rather than simply giving feedback on proposals, citizens are able to suggest alternatives and additions to those proposals. In Dublin, an example is Fix-Your-Street,[6] wherein citizens can use an online tool to report the location of issues that need to be addressed (such as potholes, graffiti, broken streetlights, illegal dumping), thus suggesting an alternative work program for city workers. Smart Dublin also has appointed an advisory network of 40 key stakeholders drawn from government, companies, universities, and civil society that meets twice a year to offer constructive feedback on Dublin's 'smart city' initiatives.

A potential by-product of citizen engagement, then, is citizens and their views sliding down the scaffold to 'product', for example through user-testing and feedback which can often occur without citizens being aware that it is occurring. For example, in the Smart Stadium[7] Internet of Things prototypes are being trialled for monitoring crowd behaviour, service performance, and stadium management. Here, feedback is given passively through mere presence and action. For Arnstein, the solution to these tokenistic forms of participation was what she termed 'citizen power'.

Citizen power

At the top of Arnstein's ladder are what she argued are more rewarding and representative forms of civic participation in which citizens have "increasing degrees of decision-making clout" (1969, p. 217). In 'partnership', planning and decision making are shared with agreed-upon ground rules and mechanisms for moving projects forward and resolving impasses. 'Delegated power' occurs when citizens gain the dominant decision-making authority and genuine specified powers within a co-shared initiative. Finally, 'citizen control' happens when citizens are fully in charge of the policy and managerial aspects of a program or institution and "can negotiate the conditions under which 'outsiders' may change them" (1969, p. 223).

In Dublin, it is difficult to identify an example of 'partnership' or 'delegated power' where initiatives are co-owned and co-created, and citizens share or have the dominant decision-making authority. Usually, examples are drawn from community development initiatives that are undertaken through a partnership between a community organization and the city, but such initiatives have not yet been created with regards to the 'smart city', or are in a very pioneering stage in a bunch of cases. Where co-creation does occur, it is usually through short-term hackathons or civic hacking/Living Labs projects. There have been a number of such hackathons sponsored by the Dublin local authorities, along with corporate partners such as IBM and Intel, with respect to using the city's open data and producing 'smart city' applications. While citizens who attend are free to produce whatever application they desire, the event is very much owned and run by the sponsors, who frame the event aims and provide space, mentors, and guidance (Perng & Kitchin,

2016). In the Dublin case, a number of prototypes have been further developed post-event into commercial enterprises, such as Building Eye.[8] From this perspective, hackathons are a means to kindle and maintain business-led urban development and entrepreneurial urban governance (Perng, Kitchin, & Evans, 2016), rather than producing citizen- or community-led 'smart city' solutions.

Concluding remarks

The chapter has unpacked how citizens are framed within 'citizen-centric' initiatives through the myriad of social and legal positions they occupy in relation to networked and algorithm-led technologies implemented in cities. In so doing, it constructed a much fuller than previously documented typology of citizen roles, the forms and nature of citizen involvement, and the underlying political discourse which sustains them. As our field site of Dublin makes clear, there are numerous roles citizens play in the 'smart city'. They can experience, at the same time, different forms of empowerment and participation. To this end, the suggestion for critical scholars and stakeholders in 'smart city' initiatives is then to use 'the scaffold' (Cardullo & Kitchin, 2019a) as a heuristic tool to critically evaluate the citizen-focused nature of smart technologies and projects beyond the rhetoric offered by the 'smart city' discourse. If 'smart cities' are going to be populated by 'smart citizens', then city administrations should be seeking to shift as many of their initiatives as possible up the scaffold towards citizen engagement and citizen power.

However, the neoliberal drive through which 'smart city' initiatives have been implemented, in Dublin and elsewhere, suggests that city services are transformed through marketisation and further privatised for the benefit of corporate high-tech. With the exception of some 'citizen power' initiatives, all levels of the scaffold are consistent with neoliberal citizenship and its emphasis on personal autonomy and consumer choice: individuals performing certain roles and taking responsibility for their own life chances (entrepreneurial self), in the context of the marketisation and privatisation of services and infrastructures (retreat of the state from service provision and austerity policy). 'Smart city' technologies, the data they generate, and the analytics applied to them can have significant negative direct and indirect impact on peoples' everyday lives (Kitchin, 2016b). They also ensure that any emancipatory and progressive policy derived through the present configuration of technologies is dependent on 'systems' that inherently surveil and control. As such, there is a potentially heavy cost for the freedom and choices these technologies claim to offer, which requires careful consideration and redress (Kitchin, Cardullo, et al., 2019). In making the city 'smart', ordinary lives become imbricated in digital processes which are often networked, machinic, and indifferent to individual's will or entitlements. The current rhetoric, instead, highlights the 'smart citizen' as an individual in possession of certain digital rights and abilities, skills, and agentic power. This individualistic dimension of digital rights intersects with consumers' preferences and choices, digital labour, collection of data for business and governance purposes, and 'smart city' infrastructures.

Do we need the concept of 'smart citizen' after all? Or can this rather be trumped by 'smart city' initiatives built on a framework which enables citizenship and communitarian ownership by default, that is as a policy direction that fosters a form of 'technological sovereignty' like in Barcelona, where digital rights are considered as collective rights to be achieved within policy goals and mechanisms of governance enabled by city administrations and negotiated with the citizens (Ribera-Fumaz, 2019)? While the 'smart citizen' is, in fact, set in isolation in the narrow and instrumental landscape of neoliberal governance, mostly as a consumer with rights and responsibilities, the goal of fostering digital rights claims ought to be a collective and political effort. This collective effort cannot be rooted in corporate or institutional settings only, whether these embrace profit-oriented corporate social responsibility or the 'citizen-centric' ethos. Rather, it would be mindful of established rights and entitlements on which the digital 'element' is further applied. Progressive and municipalist experiences of citizenship, through and about digital technologies, are the focus of the second part of this book; I will look closely at the various alliances, strategies, and initiatives that cities around the world are putting in place to defend citizens' rights and to enhance their direct participation.

It is clear that significant normative work is required to rethink 'smart citizens' and to remake 'smart cities' if they are to truly become 'citizen-centric'. The normative challenge to creating truly 'citizen-centric smart cities' will be to reimagine the political economy of cities and the role citizens are to play in their conception, development, and governance. This normative challenge would involve also reframing the paternalistic and market-driven notions of 'smart citizens' towards one rooted in a form of citizenship underpinned by collective rights and entitlements for the common good.

Notes

1 A more articulate discussion on commoning is provided in the second part of the book.
2 www.smartdublin.ie
3 www.dublindashboard.ie
4 www.dublinked.ie
5 www.civiq.eu
6 http://fixyourstreet.ie
7 At Croke Park Stadium, an 80,000 seat venue owned and operated by the Gaelic Athletics Association. https://dcu.asu.edu/content/smart-stadium
8 www.buildingeye.com

5 Living Labs and the city

In the previous chapter, I highlighted the issue of 'citizen power' in relation to the roles and participation of digital citizens in the 'smart city'. At the top of the 'scaffold' (see Table 4.1) we ideally positioned projects that delegate power to the citizens, and these being able to eventually change the goals and outcomes of prescribed policy through digital technologies. Unsurprisingly, very few initiatives are able to involve citizens directly and to devolve enough power to them. To me, such configurations are difficult to achieve if the neoliberal frame is left intact. This is because co-production involves partnerships and shared ownership rather than civic paternalism, stewardship and the pursuit of private profits. Where co-creation does occur, it is usually through short-term hackathons or civic hacking and Living Labs projects. In Dublin and elsewhere, there have recently been a number of Living Labs initiatives that adopt Lo-Fi technologies, such as sensors for the monitoring of pollution. These initiatives typically work with a community of interest and are usually university- or industry-led. As this chapter will debate thoroughly, while such initiatives do involve citizens, the form and level of participation are often circumscribed. In addition, projects led by one or a handful individual initiatives are often hamstrung by decision-making processes being dominated by a 'benevolent dictator' (Ljungberg, 2000). An example of a local authority-led initiative was Dublin Beta, which trialled street-based pop-up initiatives working with local citizens, though most were low- or no-digital technology in nature (such as pop-up parks and secure bike sheds in parking bays) (Perng, 2016). In the Dublin Beta case, the project was led by a single Dublin City Council employee who drove the entire initiative.

As discussed thus far, the spin from the high-tech industry in search of new markets and profits mixes with city administrators' dreams to running the city as smoothly as possible and to saving money in the process. Under the ideological spin of neoliberal urbanism and as a result of the related austerity policies, cities have been facing measurable outputs and benchmarks of almost everything and relentless cuts in their budgets. While endogenous budgets shrink, neoliberal 'smart' urbanism aims to attract foreign direct investment, offering areas of the city as testbeds to pilot new technologies, fostering innovative indigenous start-up sectors or digital hubs, and attracting mobile 'creative' elites. Living Labs have a big role within this strategy, to the extent that the 'smart city' has emerged as the latest, tech-led phase of the entrepreneurial city (Hollands, 2008; Shelton et al., 2015). Eventually, with the creation of smart districts and private enclaves – 'cool' places for the new technology-prone elite of young professionals and more afflu-ent buyers – intra-city competition reinforces the speculative approach to housing (the idea that urban land and dwellings have, first and foremost, exchange rather than use value).

Conversely, the mission to produce 'smart cities' has been critiqued for being overly technocratic and top-down in orientation, serving the interests of states and corporations more than they do those of citizens (Greenfield, 2013; Kitchin, 2014b). According to these critiques, 'smart city' initiatives are underpinned by a neoliberal conception of citizenship that favours consumption choice and individual autonomy within a framework of constraints that prioritize market-led solutions to urban issues, reinforced through practices of stewardship (for citizens) and civic paternalism (deciding what is best for citizens) enacted by states and companies (Cardullo & Kitchin, 2019a), rather than being grounded in civil, social, and political rights and the common good (Kitchin, Cardullo, et al., 2019; McLaren & Agyeman, 2015). Living Labs have been promoted as a viable response to these critiques, too, with the developers, promoters, and deployers of 'smart city' technologies and initiatives seeking to reposition them as being citizen- or community-centric.

This is the link critical geography scholarship starts to establish at the heart of the 'smart city'. Living Labs initiatives, digital start-ups, hackathons, smart dis-tricts, and testbed opportunities seem to enable bottom-up 'smart city' initiatives by opening these technologies to citizens and local residents, at least to the few who can meet the tags of economic, social, and technological capitals required. At the same time, though, they also appear to be the drive for transforming urban space into 'cool' techno-districts. I aim to consolidate this argument and start to evaluate the impact of the Living Labs strategy on the real estate and housing market and, ultimately, on the 'right of inhabitation': this is, in my view, a foundational issue at the heart of the broader 'right to the smart city' (Cardullo et al., 2019).

With this goal in mind, the present chapter will offer a panoramic on the Living Labs phenomenon, mapping its current configurations in relation to the trans-formation of urban space. It discusses ethical hacking and Living Labs, citizen science and crowdsourcing, and the explicit policy goal of creating smart dis-tricts and 'cool' places with a rising tag on private rents. At the end, I set the scene for a new research agenda around the Poblenou 'smart district' in Barcelona (rebranded 22@ in 2017), postulating a vicious circularity between the different circuits of capital (Harvey, 1978): this district has become the main focus of real-estate investment attraction during the post-crisis (last ten years), and has concen-trated about half of the new offers for residential and work places in the Catalan capital.[1] While the techno-led cultural appropriation of urban imaginary feeds into the development of 'smart districts', economic advantages and financial specula-tion in real estate become a drive for technology-led regeneration of cities. That is, the 'smart city' appeal for cleaner, safer, pollution-free, and efficient districts and buildings ultimately accelerates the increase in ground rent, with the related displacement effects on local residents.

Living 'Living Labs'

The Living Labs concept is generally intended as a bottom-up approach to the 'smart city', designed to increase citizens' participation and involvement in 'solving' local issues. Broadly, Living Labs pilot prototypes of new technologies or make use of Lo-Fi technologies to foster a culture of digital innovation and support community-focused civic hacking, by running various kinds of workshops and engaging with local citizens to co-create apps and devices or pilot 'smart city' initiatives at the local level.

Living Labs were born in the open design tradition of MIT's experimentation with space-aware technologies, fostering the idea that digital technologies should first be tested by their users in *"in-vivo* settings" (Dutilleul, Birrer, & Mensink, 2010). Living Labs were given a primary role in the development of the 'smart city' in 2006 when the European Commission decided to "put the user in the driver's seat" of the innovation process (EC, 2009, cited in Dutilleul et al., 2010). They are now at the forefront of the 'smart city' strategies, given their citizen-centric focus and appeal as the target of state and EU funding (Voytenko, McCormick, Evans, & Schliwa, 2016). In other words, there has been a notable shift from passive user feedback to a more active approach based on users' involvement (co-creation or participatory design). Therefore, the Living Labs approach situates the 'smart city' as a testbed for experimenting with the design and use of digital technologies *in situ*. The 'smart city' is recast in two ways. First, as being a *beta version* in need of testing through trialling, where smart infrastructures are "white-boxed" layer by layer (Corsin Jimenez, 2014). Second, as being *citizen-centric*, a more open, affordable, and democratic endeavour, developed from the bottom-up around the needs and desires of local residents, with Living Labs supplying the necessary skills and competences to citizens.

The promoters and critics of Living Labs highlight three important characteristics that enable such a vision for the 'smart city'. Firstly, Living Labs are temporality contingent, framed with respect to the temporal cycles of projects, technologies, and funding, and often run the risk of shifting a focus away "from place-making to creating temporary events" (de Lange & de Waal, 2013). This is a problem that EU funding schemes, as seen with regards to the EIP-SCC, aims at solving with the notion of 'scaling': this strategy seeks to bring forth 'best' solutions and translate successful pilots into deliverables (Cardullo & Kitchin, 2019b). Secondly, Living Labs are a context-based experience, which is difficult to replicate in the same way elsewhere (J. Clark & Shelton, 2016; Voytenko et al., 2016); therefore, replication too has become a leading goal for EU funding schemes, aiming to translate scaled technologies and policies to other locales through spreading successful pilots to other cities (Cardullo & Kitchin, 2019b). While scaling seeks to demonstrate local application, replication seeks to demonstrate generalisation and mobility; that 'smart city' initiatives proven in one place can be deployed with similar results elsewhere. It is through this process that transferable technologies, models, or 'best practice', and their circulation are established (E. McCann & Ward, 2011). Thirdly, Living Labs are intended to operate as multi-stakeholder endeavours that include local residents, acting as a counterweight to the techno-centric, top-down approach to the 'smart city' initially forwarded by big business. Ultimately, the ambition for some is that the 'smart city' will eventually boast a model of governance in which "a community assumes political and expert management over its infrastructures" (Corsin Jimenez, 2014, p. 343). We can see how the Living Labs strategy is thus crucial to successful 'smart cities': they are the Trojan horse of technology adoption, seeking to demonstrate successful application, replicability, and democratic citizen involvement.

Following on the work done with Rob Kitchin and Cesare di Feliciantonio on The Programmable City project (Cardullo et al., 2018), I would add a fourth very important goal to the successful adoption of Living Labs initiatives in the 'smart city'. Because of their contingent nature, emphasis on digital innovation and citizen 'participation', Living Labs have been one of the preferred options in order to boost investments in real estate and to reactivate urban vacancy. This has proliferated as unfinished and empty urban space following the global financial crisis (O'Callaghan & Lawton, 2016). Living Labs have offered the potential for the creation of independent (although temporary) spaces. However, just as the 'smart city' agenda has been criticized for reproducing neoliberal rationality (e.g., Hollands, 2008; Kitchin, 2015; Vanolo, 2014), Living Labs based on vacant urban sites are at risk of being co-opted into the neoliberal model of city growth, becoming a drive for real estate speculation.

In Table 5.1, I map the current configuration of Living Labs deployment presenting five leading examples: pop-up events, university-led activities, community-organised venues/activities, citizen sensing and crowdsourcing, and tech-led regeneration initiatives (adapted from Cardullo et al., 2018). The table is a heuristic, an entry point in the debate which starts to link 'smart city' adoption with participatory and emancipatory ideals, on the one hand, and the circuits of capital accumulation in cities today, on the other. Our analysis is based on a patchwork of different approaches at different times by each author (Rob Kitchin, Cesare Di Feliciantonio, and myself): interviews with observation of many Living Labs projects, hackathons,[2] and social centres' activism in Dublin, London, and Modena; desk-based research of secondary sources; and fieldwork concerning 'smart city' initiatives in Dublin as part of a large European-funded project that involved more than 300 interviews and participant observation by a number of team members, with many focusing on civic hacking, urban data commons, and the role of citizens in the 'smart city'.[3]

It is important to stress that these categories are not exclusive – for instance, a crowdsourced project can enable forms of communal engagement and ownership of the data produced for anti-gentrification purposes, assuming citizens have the political capital to act upon the data. Neither these categories are unique to each typology of Living Labs – for example, different Living Labs initiatives can be

Table 5.1 Living Labs typologies in relation to 'smart city' vacancy and city governance

City model	Living Labs type	Vacancy	Governance	Policy	Examples
Pop-up	Localised, community-based, artistic projects	Pop-up, communal or commercial	Active citizenship or technocratic stewardship	Creative city, cultural displacement, cultural capital, empowerment	Dublin Beta, 'fillit'
Digital Literate	Educational	Re-use, partnership	Stewardship, paternalism	Empowerment, steering, instrumental, cultural capital	Officina Emilia Coderdojo
Communal	Localised, community, artistic projects	Squat, social centre, moderate rent	Ownership, membership, participation, or stewardship	Communitarian, anti-gentrification, cultural capital, empowerment	Open Wireless Network (OWN), Tog
Crowd-sourced	Geo-located, environmental sensors	Maps, secondary data	Active citizenship or non-participation	Eco-sustainability or data-products;	OpenStreetMap (OSM), Inside Airbnb, Re-Using Dublin, AIRO mapping, Derelict Sites register
Regenerated	Smart districts, hackathons	High rent, exclusionary	Non-participation, stewardship	Displacement, gentrification, creative city	Digital Hub, Silicon Docks

Source: (adapted from Cardullo et al., 2018)

co-opted into the 'creative city' model of city growth, whether they are pop-up artistic projects or university-led experiments.

Pop-up Living Labs

The pop-up approach to temporary space-making has a long tradition in tactical urbanism for play-and-disruption – from the Situationist International since the 1950s (Bonnett, 2006) to more contemporary search for serendipity and discovery (Foth, 2016); from DIY urbanism, where participants intervene directly in projects (Till & McArdle, 2016); and hackable urbanism, in which urban space is not seen as given, but as moving elements that can be repurposed (Cardullo, 2014; Corsin Jimenez, 2014; de Lange & de Waal, 2016). These temporary interventions in city space draw on community building, civic participation, artistic intervention, alternative media practices, and guerilla urbanism.

An example of a pop-up Living Labs is the Fostering Digital Participation Project, which in 2015 set up mobile container units that travelled across Australia.[4] Another one is a Dublin City Council initiative, Dublin Beta, which has run a number of street-based pop-up initiatives working with citizens, though most are low- or no-digital tech in nature (such as pop-up parks and secure bike sheds in parking bays, new gutter run-off systems, painting street infrastructure to discourage vandalism) (Perng, 2017). Also in the city, a diverse range of pop-ups are being facilitated through a web platform, 'fillit', that aims to match vacancies and pop-up events: "for people looking for the perfect temporary venue for events, pop-ups, retail, promotions and everything in-between."[5] 'fillit' displays a business model similar to AirBnb, but with a twist: if the vacancy is listed for free, 'fillit' does not charge a fee either, because it also aims at "inspiring theatre groups, youth centres, the arts or start-ups."

Most pop-up projects involve social media platforms as an interface between participant stakeholders, technology, and places. They work well with the spatio-temporal dimension of digital interactions, which involves fast, transitory, and sometimes ephemeral connections. In fact, as de Lange and de Waal suggest, pop-up Living Labs projects embody a "shift from manipulating space to manipulating space in time" (2013). A key problem with the transitory character of a pop-up is thus its ontological nature: the vast majority of urban dwellers probably do not live, or want to live, in pop-up cities; neither might they want to dwell in temporary 'hybrid' locations (see Apostol & Antoniadis, 2020), but in actually-serviced cities. Moreover, in order to participate in pop-up Living Labs, citizens are required to be pro-active, engaged, and ready to play in any up-and-coming event. Participant citizens are assumed to be already in possession of, or are willing to receive, the cultural and social capital necessary to enable their participation. These are scarce currencies in modern urban living, more available to some people than others.

Pop-up Living Labs can aid local artists, civic hackers, and socially creative people who become stuck in the paradox of urban regeneration: "neither able to successfully collude due to art's lingering requirement for autonomy, nor to effectively opt out" (J. Slater & Iles, 2009). In this context, it is relevant to mention that Dublin independent pop-up spaces – carved in the niches of the city's rapid boom-and-bust housing policy of early 2000s (Kitchin et al., 2012) – are a recent tradition of eclectic and lively, but politically non-radical spaces (Bresnihan & Byrne, 2015).

Digitally literate Living Labs

University-led Living Labs are a model of partnership with government and industry that is currently "blossoming" (Evans & Karvonen, 2014). Typically, this scheme involves long-term educational activity with targeted groups, such as young people, students, and women. In the case of Dublin, Dublin City University has run *coderdojo* sessions since 2012, including coding, making, and games development, and also runs specific sessions for girls.[6] On its innovation campus, DCU has also partnered with TechShop, a "membership-based workshop and fabrication studio that provides access to machines, tools, software," and a "community of creative people, classes, workshops, instruction and meet-ups for digital and hardware innovators and entrepreneurs in Dublin."[7] The innovation campus is a refurbished former vacant space that used to be occupied by a state agency, but is now university property.

Another example of a university-led Lab was Officina Emilia (OE, Modena, 2000–2015), an action-research and museum-lab for the regeneration of competencies in the mechanical industry.[8] Its objective was "linking science, technology, engineering, mathematics and social sciences in a more effective way through the design of relationships, tools, and innovative pilot actions" (Mengoli & Russo, 2017). The Officina – which means 'sweatshop' in the Italian Operaist tradition – was connected with schools, teachers, and the SME sector, and its activity became sometimes part of local schools' curriculum. More importantly for the theme of this chapter, OE was located in a vacant industrial warehouse in the middle of Modena's Artisan Village, a place and a city with a long tradition of working-class and co-operative presence. After 15 years of intense activity, OE and its Museum-Lab had to close due to a change in policy of its funding bodies (Emilia-Romagna Region, for instance). The timeline on which the OE cycle is set offers some space for analysis. During the last few years, in fact, Italy has been at the forefront of the 'smart city' discourse (Vanolo, 2014): in particular, the city and University of Bologna, capital of Emilia-Romagna, have devolved significant funding to 'smart city' initiatives that include 'the citizen'. It is bewildering that, in a climate in which city and University struggle to start-up smart inclusive projects, a well-established Lab for the regeneration of (digital) competences is closed. The provision of free or affordable vacant premises can be crucial to the sustainability of Living Labs projects: city vacancies can be a host to all sorts of interesting projects that boost community engagement and citizens' participation, but they depend upon political will and the creation of flexible institutional tools, with 'smart city' strategy changing through the coming and going of different political alliances.

Communal Living Labs

Living Labs initiatives are sometimes medium-term interventions in local neighbourhoods that echo the traditional ethos and organization of community/social centres. This kind of Living Labs rotates usually around well-known members in a community of interest, who often act as 'benevolent dictators' in the various projects. These are ethical hackers or community advocates who provide the stewardship necessary to connect people and possess strong technical skills with respect to building and maintaining networked hardware and software applications. This sort of Living Labs, sometimes a hacker-space or art-space, is hosted in either vacant public or private space, but often seeks to maintain the characteristics of an "independent space": in both cases, rent can be a crucial factor for the sustainability of the project (Bresnihan & Byrne, 2015).[9] Typically, these Living Labs undertake a rolling set of projects that seek to address specific problems, such as Wi-Fi connectivity (Cardullo, 2017), civic apps (Perng & Kitchin, 2016), or planning applications (de Lange & de Waal, 2016). Examples of such initiatives in Dublin are Tog, a maker-space that includes digital projects, and Code for Ireland that meets monthly to develop civic apps, though it has no permanent space, its meet-ups migrating between the corporate offices of Google, Facebook, and LinkedIn (see Perng & Kitchin, 2016).

An example of this is the Open Wireless Network (OWN) in inner-city London, where wireless communication was indeed of secondary importance to the locals who participated (Cardullo, 2017). More importantly, for years OWN contributed to community-building, local knowledge exchange, and some instances of anti-gentrification activism. The case study also suggests that to make 'community' operative requires a great deal of stewardship (time and funding for maintenance, management, and investment in cultural and social capitals), contradicting the notion that technology should be the automatic interface for bottom-up 'smart city' projects. Rather, civic hackers and enlightened professionals provide the 'magic' of community relations, influencing projects' working and outcomes. This type of Living Labs is in fact based on trust, a social interfacing that is accrued over time.[10] However, time seems to be structurally lacking in projects that rely on social media to tackle local issues (see de Lange & de Waal, 2013).

In other words, the transitory character of the 'event' around which community-based activities are mobilised (planning applications, civic hacking, or artistic projects) raises questions in terms of the long-term sustainability of community relations activated through Living Labs strategy. We suggest, however, that such forms of civic hacking, with medium-term investment in a specific community of interest, might have a limited or even negative effect towards gentrification and cultural displacement (Cardullo et al., 2018).

Crowd-sourced Living Labs

A fourth type of Living Labs concerns the gathering of meaningful data via smart sensors or via citizens' reporting. The urban landscape becomes the Living Lab, with participants practising citizen science initiatives aimed at better under-standing local conditions, or being enrolled as citizen sensors. Examples of such endeavours include Sensornet,[11] which pooled together sensor data of air traffic noise in Amsterdam, and The Damp Busters that tracked dampness in Bristol homes.[12] In these cases, citizens' participation is demanded for the installation of monitoring sensors, but the experiments do not require continuous active engagement: once installed, the sensor generates data regardless, though citizens may be involved in data analysis and acting upon the data. Alternatively, citizens may be enrolled as passive citizen sensors, for example, their smartphones being tracked across a city by a sensor network to better understand footfall and move-ment patterns. The post-pandemic initiatives are big on contact tracing apps and solutions (see Kitchin, 2020). Here, the citizen is a 'data-point' who occasionally provides feedback on 'bugs' in the enabling device and gets, eventually, to a con-cluding workshop where he or she is taught how to interpret the data (Cardullo & Kitchin, 2019a).

Some crowdsourcing projects are being used to identify vacant property, rely-ing on users' inputs to generate pertinent data. Such forms of crowdsourcing usu-ally work by piping data from Google Map or Open Street Map into a mobile app, with vacant units and associated details and photos being located on the map. A couple of different crowdsourced initiatives relating to vacancy have been undertaken in Dublin. The first was initiated by a small group who walked around the city, noted vacant units and uploaded them to a dedicated Google Maps page[13] (e.g., O'Mahony & Rigney, 2016). This was then followed by Re-Using Dub-lin,[14] in which users can explore vacant sites or add any they have discovered. Other related sites include Inside Airbnb[15] that details the properties that are not in permanent use but are rented out through Airbnb lettings, the AIRO mapping module[16] on vacant housing identified in Census 2011, and Dublin City Council Derelict Sites register[17] – though these last examples are fixed and static sites, accessible to citizens, but not up-datable by them.

Somewhat ironically, one effect of crowdsourcing vacancy might be the pro-vision of fresh data in order to identify investment opportunities in the housing sector. Another problem of crowdsourcing projects is maintaining contributors' motivation and enthusiasm, with the site often lapsing into a static and out-of-date service (Dodge & Kitchin, 2013).

The regeneration Living Labs

For advocates of the 'smart city', one key reason for developing and implement-ing such initiatives is to help grow and sustain local economies, through attracting direct investment and fostering start-ups and indigenous SME. The digital econ-omy is seen as a key sector for generating new employment and 'smart' projects, a means to attract talented workers and facilitate economic activity, as well as being a new market opportunity. Digital businesses need to locate in an ecosystem of suitable office buildings with high-quality technical systems, a strong concentra-tion of business and support services, and a pool of suitable labour. One way to create these conditions is to regenerate an existing city area, one that occupies a central site near to key transport links and other business services, repurposing or replacing existing buildings. Here, Living Labs are seen as central to an ongoing process of 'modernisation', achieved by extending pioneering small-scale projects, design-focused Living Labs, and an entrepreneurial culture of open innovation, to the overall organisation of urban space and living.

In the city of Dublin there are two key sites of agglomeration, both of which are regeneration initiatives, redeveloping old, largely vacant or former industrial sites: The Digital Hub and Silicon Docks (see Cardullo et al., 2018). The former was established in 2003 and is managed by the Digital Hub Development Agency.[18] It is located to the west of the city centre in the Liberties, an area of long-standing social deprivation. The Digital Hub itself is housed in eight former buildings of the Guinness factory site and supports circa 90 companies employing between them 800 to 1,000 employees. As companies grow, they leave to find their own premises, to be replaced with new start-ups or SMEs (nearly 200 companies have been supported to date). The Digital Hub is also a key agent in local regeneration, using a public-private partnership model to redevelop and invest in local property stock, including student accommodation and restoring Georgian buildings and other industrial and brownfield sites for office space. Its ambition is to develop a vibrant digitally-driven economy in the area, but part of its remit is also to support the local community.

Silicon Docks[19] is located to the east of the city centre around Grand Canal basin and the old Dublin docks on the northern and southern side of the River Liffey. The area was initially part of the strategic development zone overseen by Dublin Docklands Development Agency (DDDA), which operated from 1997 to May 2012, when a revised Docklands SDZ (Strategic Development Zone) was created. While the original area included older, residential communities, the new Docklands SDZ's boundaries have been drawn to exclude such communities and, when completed, it is anticipated that it will include only 2,300 residential units, most of them newly built, high-end apartments (Brownill & O'Hara, 2015). The Docklands SDZ is already home to the European headquarters of many global IT companies including Google, Facebook, and LinkedIn. It has also been recently designated a 'smart district', an area-based Living Lab for trialling new smart city technologies such as sensor networks, smart lighting, smart grids, and 5G pilots (Heaphy & Pétercsák, 2016). Although dominated by large multinationals, the area is also home to numerous tech start-ups and incubator space such as Dogpatch Labs. Community-focused initiatives include Code for Ireland, though many of the participants are workers employed by companies in the area, rather than residents traditionally located nearby (that is, the participants are largely part of the gentrifying class).

The primary focus of this kind of Living Labs is on growing the digital economy and regenerating the area into a vibrant economic zone. While there are some attempts to engage with local communities through Living Labs initiatives, these are largely tokenistic to play out good corporate social responsibility, as opposed to creating a 'smart city' from the bottom up. Rather than local communities fully benefiting from economic revitalisation, the creative class are being drawn into these new digital hubs, displacing existing residents through soaring rental and property prices. As such, these areas are key active sites of gentrification where local authorities purposively seek gentrification as an ideal policy solution for urban change (Lawton & Punch, 2014).

Discussion

The first issue around the deployment of the Living Labs strategy for 'smart city' we highlighted is that of 'scaling', seeking to bring forth 'best' solutions and translate successful pilots into deliverables (Cardullo & Kitchin, 2019b). In this frame, the 'smart city' essentially takes a technological solutionist approach to solving urban issues (Kitchin, 2014a). That is, there is a presumption that all aspects of city functioning and life can be mediated or treated or optimized through technical solutions (Morozov, 2013). All that is required to solve issues such as congestion, energy consumption, emergency management of events, sub-optimal behaviour and decision-making through data-driven, are software solutions. Unsurprisingly, in our analysis of European funded 'smart city' Commitments (Cardullo & Kitchin, 2019b), we found a large number of city interfaces working through apps, dashboards, and generally real-time flow of data (public or not) aimed at 'solving' urban issues. Living Labs initiatives respond well to this demand for technology-led solutions: an array of apps and networks of sensors are proposed to cities for any kind of reason, to the point that the Mayor's Office of New Urban Mechanics in Boston (formed in 2010) has issued a handful of guidelines, written in a jargon-free language and addressed to every subject composing the 'smart city' community of interests (that is, developers, industry, designers, community advocates, researchers, academics, journalists, etc.): this explicitly recommends that prospective tenders "try to move beyond 'data', 'algorithms', and the like," to stop "sending sale people," and to avoid proposing a "smart city platform": "You might think that your technology is ready for prime-time, but we're not ready to buy it and put it up all over Boston."[20] It seems that some city leaders have had just enough of this bolstering climate of technology adoption, forced innovation, and forceful play-and-disruption.

The second issue highlighted above considers Living Labs as fostering experimentation *in-situ*, with localised and context-based initiatives, which are difficult to replicate in the same way elsewhere (J. Clark & Shelton, 2016; Voytenko et al., 2016). Thus, Living Labs are presented as community-based and localised endeavours while presuming to become a drive for innovation, bringing the contingent experience of their projects to other cities. This is a contradiction that I understand as an ideological proposition: technology is seen by its developers as neutral, with deterministic effect wherever deployed. Living Labs are generally able to maintain their local flavour, where community of interests and local residents meet (see also the following point), but are then steered by the allocation mechanism for funding, proper of super-national organisations such as the EIP-SCC: its Marketplace has been seen as a driving force in shaping the 'smart city' field with the participating cities split through a categorisation of Lighthouse and Follower cities (Cardullo & Kitchin, 2019b). In other words, rather than fostering subversive ideals of experimentation and radical engagement with local communities, smart innovation appears ultimately an exercise of replication via short-term and risk-averse finance (Lo-Fi technologies, small pilots, transient audience, repurposing of urban vacancy).

Thirdly, Living Labs are intended to operate as multi-stakeholder endeavours that include local residents, acting as a counterweight to the techno-centric and top-down approach to the 'smart city' initially forwarded by big business. In this, Living Labs are necessarily framed as community-oriented and user-friendly businesses or, at least, as open spaces where learning around 'the digital' is fostered. In the working definition by the European Network of Living Labs, these are said to be "*user-centred*, open innovation ecosystems based on a systematic user co-creation approach integrating research and innovation processes in real life communities and settings." In practice, Living Labs "place *the citizen* at the centre of innovation" (European Network of Living Labs, ENOLL, our emphases).[21] In this definition of Living Labs there is an evident slippage between the 'user-centric' model of Living Labs and its assumed 'citizen-centric' nature. At times, the slippage between citizens, users, and consumers is even more pronounced (see Cardullo & Kitchin, 2019b): the H2020-SCC call suggests as a meaningful impact that "the active participation of *consumers* must be demonstrated" (European Commission, 2016, p. 107 emphasis added).

It is in this climate of increased marketisation of citizens into consumers, users, and data-products that we need to frame citizen participation. Thus, even when 'smart city' projects herald more effective forms of active citizenship and citizen empowerment – for example, Living Labs, citizen-science, and open source software – they often do so by co-opting citizen contribution into the wider economic landscape of efficiency, environmental imperatives, and a business-driven city (Cardullo et al., 2018; Perng, Kitchin, & Mac Donncha, 2017). In the 'smart city' vision fostered by the EIP-SCC, for instance, citizens are encouraged, at best, to help provide solutions to practical issues which would respond to local and contextual situations – these are forms of placation, such as producing an app during a hackathon, or giving feedback on a development plan (Cardullo & Kitchin, 2019a). They are not encouraged to formulate or lead initiatives or propose communitarian projects – such as sharing initiatives, or urban forms of co-ownership of the common good (e.g., co-ops or shared infrastructures). Neither are applicants to funding asked to draw an alternative to the fundamental political rationalities shaping an issue, or to re-imagine a political debate. In this sense, 'citizen-focus' is often just a buzzword to soften critiques that projects are too state- or market-led and to draw funding.

While Arnstein (1969) viewed 'citizen power' as the pinnacle for creating cities that reflect the desires and aspirations of citizens, our analysis found how, in practice, bottom-up, inclusive, and empowering citizen involvement in key decision-making about cities is difficult to achieve (Cardullo & Kitchin, 2019b). In part this is because there have been few sustained grassroots attempts to create community-led 'smart city' initiatives, with communities tending to organize their activities and activism around addressing social and environmental issues through political and policy solutions rather than technological ones. In part, it is because the imperative for creating a 'smart city' is being driven by a neoliberal ideology and corporate interests that dominate the landscape and circumscribe a particular role for citizens which is highly instrumental.

While such initiatives do involve citizens, the form and level of participation is often circumscribed. There are also concerns as to the extent to which Living Labs using formerly vacant space, or being deployed in regeneration programmes, act as gateways for gentrification (Cardullo et al., 2018). The question with respect to Living Labs is whether the different forms outlined above work to create a bottom-up, citizen-led, participatory and inclusive 'smart city' – repurposing vacant property and digital technologies to the benefit of local communities – or ultimately work to serve the interests of capital and reinforce a model of technocratic governance by attracting middle-class creative and mobile workers. In this respect, Living Labs appear to feature all the main contradictions shaping social processes (including contentious ones) under neoliberalism.

Concluding remarks

In real life, participation and citizen engagement within Living Labs initiatives are often fairly limited, organised and run within a technocratic ideal of governance which implies stewardship and civic paternalist frames. Being citizen-led or citizen-engaged in the 'smart city' does not necessarily confer notions of citizenship or rights to the digital city, or automatically produce a new digital urban commons. The development and use of participative software interfaces seem, in fact, to produce an 'ideal citizen' that willingly subscribes to the idea(l)s of technological solutionism promoted by the 'smart city' discourse and acquires the cultural capital necessary to communicate or tinker with smart technologies. This is also a contentious issue with regard to participation in contact tracing experiments. While often worthy and much more preferable to top-down forms of governance, it is unlikely that Living Labs can fulfil 'citizen-centric smart city' goals, without these being explicitly rooted in notions of citizenship and community ownership, rather than citizen participation and civic paternalism. The root to this, it seems to me, is within a communal city model of citizen engagement.

Many of these Living Labs initiatives, therefore, seem to act like window-dressing activities around the 'smart city': they are hyper-visible, compared to their actual impact and effective participation, and this can be attributed to the large social media presence these initiatives have. The reliance on project participants, the unsustainability of crowdsourcing initiatives and the failure of cities to display their own vacant properties bring us to the issue of governance. Who is responsible for urban vacancy? Who is controlling the 'smart city'? To what extent can citizens impact on how space is calculated and used? And once data are collected and analysed, who has the political capital to meaningfully act upon the data? These discrepancies in the Living Labs strategy for the 'smart city' raise further questions: What kind of model of governance is operating with respect to the different forms of Living Labs individuated here? Are Living Labs really promoting a horizontal, open, and participatory 'smart city' or, rather, is their ethos rooted in pragmatic and paternalistic discourses that enact a form of civic stewardship for the 'smart citizen'? Thus, I would question whether Living Labs truly realise the bottom-up ethos of 'smart city' they promise, or whether they instead foreground an urban environment primed for the 'creative classes'? (Castelnovo, Misuraca, & Savoldelli, 2015; J. Clark & Shelton, 2016; Florida, 2003).

The following chapters provide an outlook on policies, processes, and practices for cities that, in one way or another, seem to be challenging the mainstream 'smart city' discourse and the neoliberal framework this is embedded in. Finally, Chapters 7 and 8 discuss a key issue at the heart of a participatory 'smart city': the provision of affordable Internet connections and the possibility that the public and the various communities of users involved, rather than the market, become the active agents in the delivery and maintenance of this strategic infrastructure.

Notes

1 www.lavanguardia.com/local/barcelona/20200205/473295126334/22-barcelona-distrito-tecnologico-licencias-obras.html
2 A hackers' marathon, usually lasting one day or a weekend, where programmers collaboratively code in an extreme manner.
3 http://progcity.maynoothuniversity.ie/
4 http://digitalparticipation.net.au/methodology/
5 www.fillit.ie/
6 http://coderdojodcu.com/
7 https://www.dcu.ie/news/2014/mar/s0314f.shtml
8 www.officinaemilia.unimore.it/site/home/in-english.html
9 The Dublin-based maker space TOG, for instance, charges a 45 Euro monthly membership fee, mostly to cover rent and utility bills: www.tog.ie/membership
10 A crucial issue I will return to in Chapter 9.
11 www.sensornet.nl/english/
12 www.bristolapproach.org/bristol-approach-projects/air-quality/
13 For an account see www.thejournal.ie/derelict-sites-in-dublin-get-mapped-969180-Jun2013/
14 www.reusingdublin.ie, which recently has been taken over by the homelessness charity Fr. Peter McVery Trust and is more centrally focused on housing.
15 http://bit.ly/2lHzKNJ
16 http://airo.maynoothuniversity.ie/mapping-resources/overview
17 https://data.gov.ie/dataset/derelict-site-register
18 www.thedigitalhub.com/
19 This is the colloquial name for the area, first used by the media in 2011 to refer to the concentration of high-tech companies around an old canal basin, riffing on the notion of a concentrated, Irish version of 'Silicon Valley'.
20 https://monum.github.io/playbook/#play1
21 www.openlivinglabs.eu/aboutus

Part 2

Cities on the move

An outlook on policies, processes, and practices

In the 'smart city' debate, the implementation of 'smart' technologies and, above all, the discursive emphasis city leaders and tech industry representatives have put on them, seem to have come to a maturity point. As cities around the world have been rolling out digital technologies and algorithmic-led processes and retrofitting their infrastructures, a formidable series of issues have been raised with regards to the overall re-organisation of urban life through technologies that only few can control.

On the one hand, the supposed benefits of such new technologies appear to have been overshadowed by worries around previously unknown forms of panoptical surveillance and attacks on basic privacy rights of their users. Concerns around the very logic on which the data infrastructure is built have been raised, fostering an increased demand for new forms and practices of governance, with regards to ownership, processing, and ethics of such data. These concerns take into account, among others, algorithm biases and profiling or redlining of users, commercial tracking and state surveillance, and nudging of people's behaviour. Thus, critics have highlighted how the 'smart city' might indeed have trumping effects on established human, social, and liberal rights.

On the other hand, this evolving scenario has been cast within a well-established neoliberal framework of austerity policies, privatisations, and efficiency savings. 'Smart city' ideals have been brought forth by a new alliance of city managerialism, in search of a quick fix to entrenched social issues (technological solutionism), and high-tech companies, in search of new markets or ventures through the deployment of AI and data extraction practices.

While much of Part 1 of this book has looked at the 'smart city' as an overarching meta-concept, a cultural construction, and a winning discourse underpinning a new wave of neoliberal urbanism, Part 2 attempts to unpack the diversity of responses and incipient resistance cities around the world have started putting in place in order to re-interpret, develop, and steer the 'smart city' towards a more *intelligent* city. As seen in Chapter 1, 'smart' has a few competing meanings which have been symbolically relevant for the discursive construction of the 'smart city' today: from 'algorithmic-led' processes to 'canny' and 'entrepreneurial', and eventually to 'elegant' and 'cool'. The 'smart city', therefore, is first and foremost a carefully crafted cultural construction, *forma mentis* of city managers, academic advisors, and industry representatives. This 'smartmentality' (Vanolo, 2014) is then operationalised and deployed through different forms of governance at a distance which fit well with the extractive data practice of giant high-tech companies. This scenario has left very little space to an 'intelligent city', an urban conformation of people, things, and processes that is constructed on policy goals, social justice, solidarity, and openness – not necessarily with regards to only data and software but also concerning access to infrastructure and rights to inhabitation.

In the previous chapters, through a composite mapping of 'smart city' initiatives in Dublin, we noted how Sherry Arnstein (1969, p. 216) warns against the "empty ritual" of citizen participation, which she terms "tokenism" and "non-participation." For her, effective and meaningful participation is linked to power in order to induce "significant social reform." Thus, citizens' participation would be able to change policy orientation and outcome, eventually ensuring "the benefits of affluent society" to the many. She maintains, in fact, that radically democratic participation (bottom-up) would have significant implications for the socioeconomic advancement of the "have-nots," and thus embody the potential to transform "nobodies" into "somebodies" (1969, p. 217). In other words, for Arnstein, participation through power works by reflecting an ideal of society that is more equal and just.

Marshall's (1992 [1950]) framework on citizenship works to the same ends. It is a unified concept seeking to counterweight societal division in classes, eventually "altering the patterns of social inequality" (1992, p. 44). His optimism with regards to the emancipatory goals of citizenship finds scope in challenging the unequal effects of capitalist accumulation: "economic inequality has been made more difficult by the enrichment of the universal status of citizenship" (p. 45). Although from different standpoints, Marshall and Arnstein reject a model of society founded on inequalities, social injustice, and individualism. The second part of this book takes these two authoritative frameworks as lenses for understanding citizenship ideals and roles in the 'smart city'. It argues that an alternative 'smart city' can be found only in forms other than the prevailing neoliberal paradigm. In order to achieve that, I suggest that further normative work is needed; this would be more imaginative than the window-dressing claims to the 'citizen-centric smart city' discussed thus far; and it would aim to connect the dots of an alternative 'smart city' vision and practice.

I frame this second part of the book around the concepts of the 'intelligent city' and the 'right to the smart city': these are city strategies that centre their policy on people's needs, through human-based and fundamentally democratic forms of governance, and with public or collective forms of ownership and control of infrastructures (Cardullo et al., 2019). The 'right to the smart city' concerns the possibility for people to change the overall objectives or strategies for city development; the right of all city dwellers to fully enjoy urban life with all of its services and advantages and to take direct part in the management of cities' governance (Kitchin, Cardullo, et al., 2019). This is the area that belongs to politics and political confrontation, the agonistic process of democracy at work.

It follows that an 'intelligent city' might not be technological or 'smart' at all, setting its priority in different directions: these might be technological enough or devoid of algorithmic forms of governance. Either way, I would argue that an alternative conception of ethical cities consists in moving beyond the dominant post-political framing reproduced by its epistemic community and advocacy coalitions; and, finally, in rethinking notions of 'smart' citizenship and the purposes and ideology of the 'smart city' in ways that are thoroughly political. In order to do that, we would probably need to turn the 'smart city' discourse on its head and exploit the tensions. Hollands (2008) highlights between "serving global mobile capital and ordinary citizens; attracting and retaining an elite creative class and serving other classes; and top-down, corporatized, centralized development and bottom-up, grassroots, decentralized and diffused approach" (Kitchin, 2014a, p. 2).

Thus, this second part takes a more normative approach, reasserting a renewed 'right to the city' in the digital age. In a nutshell, it re-contextualises the deployment of urban technology in the messy places of city living, suggesting that:

1 Smart technologies, as currently conceived, are instrumental to neoliberal ideas of citizenship rather than to a participated, equal, and just city. In fact, 'smart city' initiatives, pilots, and experiments tend to replicate neoliberal models of urban governance and development in which high-tech multinationals have strong power of influencing research, driving innovation, and steering policy implementation.
2 As a consequence, cities and civic society activists have started to put in place a series of drawback measures in order to protect basic rights and politically grounded urban policy, and to take back some form of control with regards to mostly automatic, autonomous, and algorithmic-led processes, technologies, and infrastructures.
3 'Citizenship' is an evolving vector in the new sociotechnological assemblages of governance, which requires interdisciplinary investigations. Thus, experiments in techno-politics and digital communitarianism ought to be activated within a broader framework of urban equality and participatory governance, based on civic and social rights and redistribution of city resources.

The possibility of reinventing the 'smart city' under novel grounds is the challenge to engage with, *in primis,* the different strategies and initiatives some cities have already adopted. Thus, the chapters in this second part present a series of case studies from cities that struggle to contain the side effects of techno-capitalism for city planning and design, organisation and management, governance of service provision, surveillance, and control of citizens and city functions.

The purpose of Chapter 6 is to provide a first outlook on this set of policy measures, legal arrangements, and market corrections to the buoyant 'smart city' discourse and practices. This outlook does not pretend to be comprehensive, for obvious reasons, but it aims at foregrounding heuristics on the evolving landscape of governance and implementation of the 'smart city' along three crucial policy issues: data ethics, measures that contain digital platforms, and 'smart city' adoption.

Chapter 7 debates the possibilities and the limits of a public Internet infrastructure which is municipalised and locally controlled, providing examples of more democratic governance for this strategic infrastructure. It is based mostly on secondary sources collected in the last few years. The chapter argues for a different approach to city governance with a degree of 'meaningful' participation, citizen control, and access to data and communication infrastructures.

Chapter 8 delineates issues and perspectives for the urban commons and commoning, and it starts asking: What kind of trust, collective intelligence, and forms of capital circulate between digital infrastructures, citizens, and the city? The chapter makes a positional argument for a 'smart approach' to the commons, taking forward the issue of municipalism and urban commons and advocating for the city as a crucial stakeholder in creating and maintaining urban commons in the 'intelligent city'. It revisits the ethnographic account of a community wireless network in inner city London, questioning determinants in the bottom-up 'smart city' strategy: 'stewardship' and 'rent'. The chapter argues such initiatives can foster a high degree of trust and social/political capital, but questions their levels of stewardship and inclusion.

The concluding Chapter 9 suggests a novel and alternative version of the city yet to come, the 'intelligent city' where hybrid notions of citizenship are reconfigured along civil and social rights, active and meaningful participation, and redistribution of socially accrued benefits for the common good. It ends opening to the author's research to come, which aims at joining the dots from digital hubs, start-ups, smart districts, and the 'techno-cool' with circuits of capitalist accumulation in cities today. It makes a link between the novel technology-led push to regeneration and the well-studied phenomenon of gentrification. This is the process through which local residents and urban poor get, sooner or later, displaced by a process of urbanisation that privileges private investments in the housing market and novel lifestyles.

6 Provincialising the 'smart city'

Critical research in Urban Geography and Planning has started to show the wide gap between the premises on which the smart technologies deployment is based and the reality of its implementation. This research stresses that cities are complex and opaque entities subject to a myriad of local determinants, forms of governance, political tensions, and that, sometimes, their policy might be steered by relentless activism from community and interest groups. Thinking that technology can be implemented and utilised in the same way everywhere appears, therefore, an ideological and deterministic proposition which has been affecting the mainstream thinking around the 'smart city'. This proposition is alimented by transnational advocacy coalitions from industry, city, and academia, and it has become a branding tool for the expansion of markets for AI and algorithm-led technology.

According to Joss (in Karvonen et al., 2018, p. xvii), the 'smart city' has become an elusive and vague concept: a "floating signifier" which holds great usefulness for policy debate since it enables generic positive futures. The winning assumptions of the 'smart city' framework, as seen in the first part of this book, can be sketched along these four determinants:

1 *managing complexity:* the algorithmic rationality behind the modelling of practically everything can bring together vast networks of nodes, as well as the tools for creating, managing, and navigating such networks;
2 *technological determinism:* interventions in such a (complex) 'system of systems' will eventually bring forth expected results which are to be true in the same way anywhere;
3 *technological solutionism:* cities should implement such software and hardware assemblages in order to achieve systemic solutions to their long-standing problems;
4 *post-political condition*: given the overwhelming presence of real-time Big Data and AI responses, policy must now answer the question "How to?" rather than "What if?" This leads to depoliticised and technocratic responses, a never ending series of enabling conditions: for instance by forecasting on real-time data, systems tend to optimise an already present traffic congestion (with apps, dashboards, traffic lights synchronisation, CCTV, and so on),

asking *how to* divert traffic, rather than asking *if* we actually want a certain level of traffic congestion in the first place, and take backcasting policy measures (e.g., cycle lanes, affordable and reliable public transportation, bonus/malus rewards, etc.) (see Kitchin, 2019b).

It follows that a process of provincialisation of the 'smart city' concept is due, and this is the task the second part of this book attempts. 'Provincialisation' is about making claims different from those master narratives of mainstream global urbanism that are universalist (or worlding); for instance, those embraced in the epistemological fundamentals of neoclassical theory: the imperative of city growth and the quantification of everything possible (Charnock, March, & Ribera-Fumaz, 2019). Therefore, "provincializing global urbanism means identifying and empowering new *loci* of enunciation from which to speak back against, thereby contesting mainstream global urbanism" (Sheppard, Leitner, & Maringanti, 2013, p. 895). Instead of presenting smart urbanism and its policy in a fairly homogeneous manner, this approach suggests focusing on the geographical arrangements of such configurations. This kind of research would be attentive to the context in which the 'smart city' discourse is deployed, looking at the emergent local practices, with detailed descriptions grounded in empirical work and analysis of "Actually Existing Smart City" (Shelton et al., 2015).

Different urban governance contexts result in a variety of patterns for citizen involvement in 'smart city' initiatives. Cities are not passive receivers of conditions dictated by markets and large companies; under certain circumstances, they can resist market forces and, within limits, actively shape their economic choice. Cities are in fact places, territories and bounded administrative arrangements of scale, in constant dialogue with their administrative, political, and economic relations. These, in turn, can be part of city life or external to them, maybe depending on transnational discourses and policy, or far-reaching lobbies and interests. In addition, complexity is part of the life of cities: the critical mass of people, labour, and capital that are present in cities represents an additional 'thick' layer to be taken into account when studying technologies and cities. These loose and changing elements of urban life are linked together in many different ways, in economic needs or political demands; sometimes, in social relations of affect; sometimes, by networks of infrastructures which deliver everyday basic services and utilities, or connect and make them able to communicate with each other.

Cities are therefore complex and opaque, and subject to a myriad of local determinants, forms of governance, and technological adaptations at different scales and, sometimes, their policy is steered by the relentless activism from communities and interest groups. Thus, talking about 'modes' or 'forms' of governance refers to the ways in which political institutions in each city are linked together by informal and structured arrangements and practices (Rhodes, 1996). These political and administrative relationships determine how cities are governed – they shape objectives and goals for the city, as well as the nature of interaction among government and local actors, including citizens. It follows that cities are much more than an interrelated 'system of systems' (Batty, 2017), a set of quantifiable relations whose flows of resources, outputs, and information are discernible and, thus, manageable or predictable. This way of thinking about cities confers a blind trust in the epistemic community of IT architects, software engineers, data analysts, and model builders to make visible, control, and predict the underlying relations which connect all those loose elements of city living.

Recent critical research has started showing that the 'smart city' phenomenon is geographically diverse. For instance, Karvonen et al. (2018) present a collection of case studies that foregrounds some of the key issues highlighted in the first part of this book as diffused; however, they illustrate also the presence of important geographical difference which takes into account the conceptualisation, implementation, contestation, and practices of being a 'smart city'. Empirical studies (e.g., Charnock et al., 2019; de Lange & de Waal, 2013; McFarlane & Söderström, 2017; Trencher, 2019) show that some 'smart cities' around the world are adopting a more human-centred agenda and are moving away from ambitions of using digital technologies and data as tools to help improve citizen lifestyles and tackle social problems. For instance, Trencher's (2019) study helps to better articulate the distinction between different 'models'. It acknowledges the complexity of the 'smart city' phenomenon, with some initiatives deviating from the neoliberal 'model', which Trencher calls "smart city 1.0." Cities, he suggests, present "collective issues affecting multiple stakeholders" for which forms of collaborative governance are required, involving diverse societal players who share a common agenda to improve local conditions. However, while Trencher (p. 120) sees "digital and data literacy of local residents and professionals" as a crucial gateway to this, I would suggest that forms of governance are contested and might evolve towards trajectories that are not to remain necessarily 'smart' or technology-intensive. As I have argued, the imaginary of the 'smart city' is fundamentally ideological, in that it is a taken-for-granted starting point for adopting new technologies, rather than the 'smart city' *itself* being questioned as a political choice.

Some cities around the world have started challenging current mainstream ideas of urban growth and entrepreneurial urbanism and embracing 'alternative' visions of technology deployed through their own urban fabric. They can be divided between those who wish to comply with innovative (especially European) legislations on digital rights and privacy, and others attempting to reconceptualise the 'smart city' concept altogether in order to present a more coherent alternative to the neoliberal one. I have given different denominations to such cities: for example, 'the democratic city' to Barcelona, 'the nearly-inclusive city' to New York, or 'the socially sustainable city' to Medellin, trying to encapsulate this variety of visions and frameworks. These denominations are not conclusive or exhaustive: rather, they aim to represent a trajectory and cultural change in the city policy. In this respect, I would highlight NYC's focus on digital inclusivity, or Medellin's experiments with socially sustainable urbanism. Amsterdam seems to have a more pronounced interest in preserving digital rights through open access and data, for instance using its annual procurement budget of €2.1 billion to help guarantee good terms for data privacy. Finally, Barcelona attempts to change the foundation of techno-capitalism and reorganise the ownership and management of city technologies, thus moving away from a compliant framework to a more alternative and imaginative one.

What is more, some cities around the world have joined in coalitions and alliances that aim, at least, to come to terms with the idea that technology cannot fix everything everywhere. Amsterdam, Barcelona, and New York together launched the Cities Coalition for Digital Rights in partnership with UN-Habitat, a United Nations program to support urban development. Cities who join the coalition agree to a declaration of just five principles that centre on respect for privacy and human rights in the use of the Internet and related technologies. This attempt is based on a broad and political application of citizens' rights, on a deeper understanding of technology as sociotechnical assemblage, and on a set of formulas that privilege bottom-up processes and city alliances with the commons and civil society rather than the market. Whether these various initiatives and, sometimes, pilots are successful or not is less relevant to my research: some of these initiatives will take some time before an assessment can be made. The aim here is rather to evaluate the overall framework and the policy attempts at changing the neoliberal orientation of the 'smart city' discourse, or steering the 'smart city' debate, or fostering solutions to urban problems that are fair, unbiased, and citizen-oriented.

The process has undergone some critique for its contradictory or hybrid organisation, being vehiculated through existing 'smart city' avenues and advocacy coalitions. For instance, Charnock et al. (2019) argue for adopting the 'rebel city' concept as a heuristic for reflection on the recent repurposing of the 'Barcelona model' of urban transformation. In fact, this 'model' has followed swinging positions in the political alliances, choices, and discourse. More importantly, a composite picture of resistance emerges from mapping 'actually existing smart cities', although this too remains difficult to enclose into an alternative 'model'. Rather than delving into definitions, such as 'citizen-centric smart city' (e.g., EIP-SCC) or 'Smart City 2.0' (Trencher, 2019), the suggestion would be to set the basic principles around which very different cities can talk to each other and share experiences, practices, and also investments; that is, setting the trajectory in urban policy that will move towards an 'alternative' smart city (or, as I prefer, an 'intelligent city').

It follows that a process of provincialisation of the 'smart city' is not only due and necessary, but it presents some edge points and tensions, too. In the following three sections, I give an outlook to this emerging landscape, focusing around data ethics; measures cities start putting in place to contain the side effects of digital platforms; and the overall concept of 'smart city' adoption.

Data ethics

As seen in Chapter 3, two ethical frameworks emerge through which the enactment of rights can be framed and nurtured in the 'smart city': the procedural and compliance model *vis-a-vis* the normative and ideological one. These two 'schools' of doing and enacting ethics can be also framed as post-political *vis-a-vis* agonistic process. Deontological and utilitarian theoretical frameworks for ethics pose mostly procedural and instrumental questions, such as: "how to best adopt technologies already decided?"; rather than asking ontological questions, like: "Is this technology appropriate or founded on human rights in the first place?" Ethicists in residence say: "We have these robots and AI, so *how* can this be made less biased and privacy-aware?"; while Legislators would argue for a fair assessment of the dangers and lock-ins of technological development, an analysis which concerns *alternative* futures to the prescribed one, for example, the 'smart city'.

Thus, the first model presents an increasing preoccupation with ethics as a guide to 'do good' by way of correcting glitches in the 'smart city' systems (at least, those that become evident) without changing the *status quo* or the fundamental structures of society, such as the market-led impulse for creating the 'smart city'. The second model, the normative and ideological one, would instead look at the power relations that make such an 'issue' possible in the first place, for example, structural inequalities in society, data ownership, and effective forms of citizens' participation. For instance, danah boyd declares, addressing the Electronic Frontier Foundation Pioneering Award on issues of institutional sexism and gender discrimination within the high-tech industry: *"The goal shouldn't be to avoid being evil; it should be to actively do good. But it's not enough to say that we're going to do good; we need to collectively define – and hold each other to – shared values and standards."*[1]

The data infrastructure emerging in cities around the world has rarely been planned, designed, and implemented against a rights-based framework. The recent emphasis on Data Ethics, especially from Industry and city managers, has therefore raised some suspicions in that it appears more as a procedural and instrumental exercise, which aims at avoiding a perceived 'barrier' to the adoption of technologies, rather than a trope able to change the current discourse around the 'smart city'.

Amsterdam: The open data city

TADA is a participatory manifesto created with local business, academia, and residents. It has been followed so far by 76 government authorities, companies, and other organizations from different regions who share their ambitions to shape a responsible digital city, along the policy routes of data minimization, open by default, privacy by design, and a ban on Wi-Fi tracking.

TADA, 'data disclosed' manifesto,[2] has formulated six principles: to be inclusive; people's control on data; right to be forgotten; citizen-led planning; open; and data commons. TADA principles are not much different than established ethical values. However, the challenge the organisation decided to face is how to apply them to the digital domain. In this sense, "TADA is a movement," an interdisciplinary and operational effort based on a set of solid principles and rights. That makes data ethics and principles not an exact science, but it requires stakeholders and individuals to use their moral compass at each step. "That takes time: time to think, time to engage in dialogue, time to ask each other questions." Ethical principles, such as 'the human factor' and 'from everyone – for everyone' are soft values that need to be reinterpreted time and again: "Ethics cannot be outsourced," says Tessa Wernink who partners in TADA: "It is important for everyone to think about that in their work."[3]

According to Amsterdam's city director, it is the scope of using data that needs to be transparent in any case: "There might be a contentious issue, that it feels like it's right on the edge of what's acceptable. But that is exactly what *political debate* is for."[4] This is where TADA stands out, in my opinion, in the notion that a political subject has to take a stand on such issues that are inherently political. In addition, OpenCity[5] Amsterdam provides easy-to-use digital participation tools to include residents' voices, for instance in co-creating the design of public space (de Lange & de Waal, 2016).

Amsterdam is also the founding member, with NYC and Barcelona, of Cities for Digital Rights, the coalition of over 100 municipalities that wants to put citizens' right first and treat data as a civic infrastructure for the commons and the public good for the many; and to guarantee that the focus is shifting to people and their rights, interests, and preferences, rather than technology for the sake of it. For their attentive and holistic approach to data infrastructure, I would give Amsterdam the engaging title of: "the open data city."

Measures that contain digital platforms

The rise of digital platforms in the mobility, consumer, and accommodation sectors has been relentless in the last decade. The digital platform economy is said to have the potential of not only realising smart mobility and accommodation in modern cities, but also of radically disrupting the nature of work and the structure of the economy (Kenney & Zysman, 2015). Consumers and producers of goods or users and providers of services are now meeting on and exchanging through these platforms, allowing new ways to distribute resources, remotely, through apps, games, and social media. As a matter of fact, digital platforms have been powered by high-tech firms, such as Facebook, Google, Amazon, or Uber, all of which leverage advanced and secretive algorithms, big data, big data analytics, machine learning, artificial intelligence (AI), and cloud computing.

The corporations behind digital platforms now occupy the top positions as the world's largest capitalisations. Successful stories of start-ups and much bigger tech companies gaining important shares of markets, however, hide the way in which transnational high-tech finds a law loophole and disrupts the market as fast as they can to reap profits, while using such profits to lobby against regulation. These successful narratives hide, moreover, the support given by the public sector in terms of infrastructures and subsidies (Lazonick & Mazzucato, 2013). Thanks to law loopholes, they proved to be very disruptive in various economic sectors such as transportation and lodging; they have transformed the press, through the promise of on-time, always on, 'citizen journalism'; and they are penetrating education and health sectors very rapidly, despite the increased sensitivity of the data treated in those sectors, and the recent coronavirus crisis has accelerate this process enormously. The outcome has been the corporatisation of public services (such as transport) or markets increasingly dominated by private players (such as Airbnb operations or Amazon logistics) (Kitchin, Graham, et al., 2019). Platform economies are all related to their capacity to treat information, streams of data, and consumers' profiles and preferences, and to predict and steer people's behaviours.

For instance, thanks to Airbnb, home sharing has become a more popular form of tourism letting, leading some professional landlords to withdraw houses and apartments that would normally be rented on a long-term basis in order to rent them out as short-term lets on the popular platform. In addition to the obvious competition with the lodging industry, the transformation of one's own home into a holiday home initiates and promotes various negative externalities, especially on the housing market and in the local community – such as short-term rentals invading neighbourhoods, creating potential tax compliance challenges, and adding new zoning. The overall inflationary effects on residential rentals are felt wide, too.

As a consequence of platform economy side-effects, many cities have started introducing various forms of control, pieces of legislation, new taxes, and restrictions. With regard to the Airbnb example, an agreement on the number of rental days for the apartment as a holiday home, known as a 'cap', serves usually as the basis for policy regulation. This is the case of Barcelona which introduced a city-approved license and forced the platform to give city officials access to users' ID and rental data. New Orleans too managed to impose on Airbnb the obligation to share with the city their rental data, but this is more of an exception than the usual policy direction, as we will see from the report below with regards to Airbnb Ireland.[6] New Orleans has also banned the platform rentals by zoning (in tourist areas altogether, such as the French Quarter), or the short-term rentals on an entire property – meaning, any property not occupied by its owner. The control and enforcement of these approaches, however, remains difficult and often inadequate. Some cities started to experiment with their own ethics-proof renting platform: one example is FairBnB, an ethical coop which promises to foster transparency and legality in the sharing rental market, with their 'one host, one home' policy and payment of taxes to the local authority; 50% of the commission is to be used to fund community projects; and opening up of the organisation ranks to the community of hosts and services who can become members of the coop.

Cities that are taking a stance against, or attempting to regulate, Airbnb are acting on the principle that 'houses are for homes, not hotels'. Further, housing activists who are often behind such policies (as for Barcelona) have questioned the rule of the market as free arbitrator: houses should be evaluated in relation to their use value first, rather than exchange value. Finally, such cities and, sometimes, social movements have reasserted the need for a right-based framework over individual gains: houses are shelters where humans can fully exert their right to inhabitation, their basic 'right to the city'.

Airbnb Ireland

Inside Airbnb and Irish Housing Network have released a first major report on the impact Airbnb is having on the Irish rental market, at a time when this is affected by a deepening housing and homeless crisis (Flanagan, Connolly, Brown, & Cox, 2019). This situation is particularly felt in Dublin which has experienced "boom, catastrophic bust, trenchant period of austerity and crisis, and recovery in the property market that has resulted in a new crisis of housing affordability and associated homelessness" (O'Callaghan, Feliciantonio, & Byrne, 2017, p. 87). The details of such a crisis are painful: according to the Dublin Region Homeless Executive, in November 2018 there were 1,296 homeless families living in 'emergency accommodation' such as hotels and homeless accommodations, with 2,816 children;[7] overall, more than 10,000 homeless people were counted in Ireland in 2019.[8]

And yet, around 5,000 'entire homes' on Airbnb Ireland platform appear to have been rented out for more than 90 days – a good indicator that the host is not living in the property, according to a new legislation (2018) which labels such occurrence as 'commercial venture'. These homes account for more than half of the listings with a staggering 74% of estimated revenue of the entire Airbnb Ireland, the report suggests (Flanagan et al., 2019). A commercial Airbnb listing is when a listing does not represent a temporary use of a primary residence. Triangulating different data, the researchers estimate than about half of the listings in Ireland are from commercial portfolios controlled by property investors and managers, and their impact is geographically relevant: in Galway city centre, for instance, over 7% of private rentals are commercial Airbnb listings; in the tourist locations of South and West Kerry, it is over 50% of private rentals.

Thus, Airbnb is not a small issue in Ireland. Until now, Airbnb has been "unlegislated and unregulated" in Ireland and the new regulation is "not fit to task," the new report suggests, proposing a Platform Compliance framework for regulation: "Platform Compliance would simply require Internet platforms to only advertise listings that have a required permit from a local agency" (Flanagan et al., 2019). However, Airbnb does not release data to Irish governing bodies on who is listing houses and it is actively lobbying local and national governments in order to stop unfavourable rulings. And this is where things get interesting from an ethical perspective: *pace* corporate social responsibility, Airbnb questions the validity of Inside Airbnb data; but since the digital platform doesn't share its own data publicly, the only available information comes from third-party estimates. The actual extent of illegally repurposed housing units into holiday homes is thus unknown and can only be estimated with the support of the platform operators themselves, who are usually uncooperative in this respect.

Indeed, the new legislation trying to regulate the effects of the platform in Ireland implies platform compliance, that is, Airbnb actively stopping commercial listings and making its data available for inspection. And this is not happening.

The Gig economy

A large part of the platform economy is the provision of services to individual consumers, whether food or transport. The platform economy has also been called the 'Gig economy' because it largely consists of private hire drivers, couriers, and other low-paid workers, who are infamously classified as "independent contractors". Whilst it's true that most private hire drivers and couriers are self-employed, this has now been foregrounded as a strategy to deprive them of employment rights to which they are legally entitled. What work tribunals in the UK suggest, instead, is that they are "limb workers", a type of self-employed person who is still entitled to rights like minimum wage, paid holidays, pensions, protection from discrimination, and more (see Scholz, 2017).

Pace corporate social responsibility, on this ground, too, platform companies are uncooperative with cities and unions, actively obstructing attempts to regulate this growing sector of urban economies. For example, Uber has appealed the ruling which allows its minimum-waged drivers to access all platform data concerning *their own* performance, including detailed GPS records, trips, earnings, full log-off/on times, and management actions. Some drivers have been fired – by an automated decision – for cancelling or not accepting work, and they've been given no right of appeal. Revelations on working conditions for Amazon employees are telling about the lack of welfare at work and workers' inability to follow on the pace at which robots work. While compiling this report, I have been following on Twitter the Deliveroo drivers' strike in Peckham, Camberwell, Bermondsey, Brixton, Clapham, Balham, and other areas in South London, where I lived for the last 20 years. They are fighting for better pay and conditions. As one of the Independent Workers' Union of Great Britain secretaries (IWGB_CLB), and Deliveroo rider, writes: "Worst I have seen working for Deliveroo is people earning £2 an hour after expenses . . . it is exploitative and precarious." From the perspective of the ethics of AI or data, Deliveroo will eventually calculate a break in between deliveries, or treat employers' data in a sensitive manner; from the normative framework point of view, those employees are still paid below the minimum wage, and their welfare rights have been severely infringed upon.

Gig economy and sharing economy are potentially intelligent ways of providing income and services in places where more established companies or cities fail, but there is a fundamental bias at the root of this idea: *both* the end user *and* the service provider are organised, disciplined, and rewarded by *the same* platform powered by an app, that is a series of algorithms whose working, processing, and data extraction abilities are hidden to any scrutiny. It is *the same* company which also extracts the valuable data to make sure the market runs smoothly to their own advantage: digital platforms have widely exploited new types of feedback loops on information systems, probably more than any other industry in the past. Essentially, these loops are based on the following frame: the more users a platform has, the more information can be collected; thus, more users are keen to join the platform. Obviously, platform companies can decide *prices* and allocate services according to the inscrutable laws of their highly secretive algorithms: *the invisible hand of the market,* as Hayek suggested (see Fuller, 2017). Unfortunately, only a few can see this hand and steer it to their own advantage.

In sum, there is a great need for cities and governments to invent, pilot, and share more regulations and pieces of legislation, sometimes with painful effects on the above-mentioned companies/platforms. It is becoming more common that cities implement caps, licences, new taxation, or even ban altogether platform economies from their territory. It is also becoming more common that cities join their strengths in coalitions in order to act against the powerful transnational high-tech companies behind the platform economy, in order to have more contractual power and to avail such measures together. The Sharing Cities movement is a case of this, an alliance born in 2016 and now meeting regularly in order to better understand the disruptions platform economies are bringing to cities and citizens, and what innovative measures can be taken to meet the challenges and opportunities cities face. One main goal of the coalition is to regulate Airbnb, in order to limit negative impacts on the availability of rental units for local residents and the inflationary effects on the housing market in general: it is not a coincidence that the first city to join Sharing City coalition was Amsterdam in 2015, which is also the first municipality in the world to develop regulations around Airbnb.

Uber too has faced many restrictions and increased regulation in many cities. For Barcelona city officials, the Silicon-Valley company does not own any car in the city and it is in breach of fair competition with local taxi drivers; therefore, the city imposed a series of restrictions which initially provoked the company to suspend its activity in the city: drivers were asked to wait 15 minutes to travel between pickups and to hide their car location during the booking process. Barcelona has been accused of breaching free market, thus lobbying and putting the interests of their taxi drivers before the opportunities offered by the high-tech giant to potential drivers (in terms of 'jobs') and benefits to their customers (in terms of reduced fares). In 2017, however, the European Court of Justice ruled that Uber is effectively a transportation firm and not simply a digital app, thus forcing the company to deal more closely with local governments which set transportation rules and licensing requirements.

The truth is that very few predicted what was coming when digital platforms started to take over cities: some were lured into the hypes of 'technological solutionism', where a cash-strapped city budget would find a ready-to-use and algorithm-regulated solution to deep-seated urban issues, such as pollution, congestion, or crime. Some commentators from the technology columns of their digital magazines suggest, with the vigour of an impenetrable faith, that the great achievement of algorithmic-led apps which allocate goods and services in the sharing economy is the emergence of a new form of trust, horizontally established between the two sides of this market, say, between the drivers and their passengers who evaluate each other with ratings and feedback. This new form of trust would *de facto* function as a replacement of the vertical trust provided by institutions at large, including their regulatory and sanctioning systems. In other words, the algorithm-regulated trust is thought to be working in the same way as the invisible hands of the market, allocating goods and satisfying preferences or optimising results in the most logical way, the logic of the market. Platform economies' pledge is to transform precarious working lives, often at the bottom of the social order, into smart and feral entrepreneurs. But the real disruption this model is bringing about is against the social pact between capital and labour, which led to the modern welfare state. For some, these relations are simply part of the rules of the game within a market economy, at times called the 'labour market', at other times the 'cost of labour'. For others, this is more to do with the breaking of those ties on which liberal democracies have been founded.

While some commentators think that capitalism is all about vertical concentration in the hands of few rather than working *also* as de-centred nodes for the benefit of the same few, for critics this de-centring gives room to a lack of accountability for firms and the impossibility for users of lodging complaints or steering the policy direction of the digital platform. Service users are framed mostly as 'data-points' and 'consumers' moving between the categories of 'non-participation' and 'choice' (Cardullo & Kitchin, 2019a). Social trust, moreover, is a slow and mostly face-to-face process that is built over time, upon frequency, social bonds, and culture. To this crucial difference, I will return in Chapter 8 where I delve into how the cooperative movement, a sector traditionally devoted to sharing and provision of services to the person, is responding to the buoyant platform-based gig economy.

'Smart City' adoption

As discussed throughout Part 1 of the book, the 'smart city' is a solid and growing discourse sustained transnationally by advocacy coalitions and epistemic communities. These forms of consultancy are often linked to academia and industry, whose strategies for funding research aim at sharing and scaling up the 'smart city' vision: by way of funding new technocrat posts, seeding and demonstrating the potential of the 'smart city' vision, and fostering digital learning. Such epistemic communities and advocacy coalitions seek to corral and shape the technology adoption field by eventually promoting an algorithm-led model of urban growth. This 'model' favours mechanisms for the adjustments of opportunities, such as digital platforms which allocate resources, display pre-packaged solutions for various stakeholders, and favour exchanges within already determined boundaries of cooperation. Industry and academia policy agenda then translate these ideals into a programme, the 'smart city' adoption strategy, which currently includes further splintering of infrastructural provisions and a general sense that the market rather than the state can allocate common resources more efficiently, with privatisation and outsourcing of services. Thus, the problem with mainstream 'smart city' adoption is that this embeds sectoral assemblages of neoliberal governance able to form and move swiftly through diverse cities across Europe and at multiple scales, promoting a model of participation that is rooted in pragmatic, instrumental, and paternalistic discourses and practices. This multi-scalar perspective on neoliberal governance is made more relevant by the current phase of austerity politics that forces cities, deprived of autonomous spending capacity, to compete against each other in order to attract supra-national investments (Peck, 2012), in the form of creative classes and digital hubs of super-connectivity where testbed implementation has been forced upon cities and their inhabitants. Such a flow of investments, both domestic and international, has to be maintained through policy agreements, a climate of optimism with regard to 'smart city' solutions, and a sense that cities cannot lag behind: this is the purpose of 'smartmentality', a cultural and political drive to technological innovation and to the active creation of markets for the new technology.

Other areas of intervention, for cities willing to fight negative side-effects of digital capitalism and its platform-based economy, are therefore public procurement and the capability of negotiating legally binding contracts through the policy tools of licensing and planning. Individual cities can put up a notable negotiating power or they can join in larger coalitions when dealing with high-tech providers and city-wide platforms (as seen in the Airbnb debate). Despite the alleged lack of technical competency, cities are already putting in place ethics-proof contracts: for instance, Internet provision contracts can include stringent clauses in terms of privacy by design, openness of data, and related digital rights matters. When the Federal Communications Commission decided at the end of 2017 to wave Net Neutrality principles in the United States deliberating in favour of Big Cable monopolies, mayors of many cities pledged to only contract with broadband companies that would agree to abide by Net Neutrality rules. "Over 80,000 letters were sent, and the coalition grew to over 200 cities in 41 states representing more than 25 million people" (McDermott, 2019, p. 19). New York City, for instance, managed to obtain a very favourable deal from Big Cable for the supply of the Internet to its pilot project in Queensland Estate, which gives free broadband access to all residents of the housing complex, although capped at 25Mbs speed which, at the present, is the minimum standard for defining broadband connection (the case is discussed in more detail in the next chapter).

Therefore, the 'smart city' is a political and cultural construction which depends heavily for its adoption from alliances, coalitions, lobbying, and pressure groups from different stakeholders and civic society groups. This is evident, for instance, in the recent journey of Barcelona 'smart city' (Charnock et al., 2019; March & Ribera-Fumaz, 2016; Ribera-Fumaz, 2019).

Barcelona: the democratic city

Under a right-wing neoliberal government, Barcelona became the poster child for the 'smart city' through its various initiatives and aggressive self-promotion and hosting of the Smart City Expo and World Congress (SCEWC). The 'smart city' discourse appeared so solid that the massive anti-austerity protests in 2011, which challenged the economic crisis and also the role of local government in urban growth and development, were unable to shift the agenda of Barcelona government. However, the city's vast educated, politicised, and networked population was hit fiercely by the financial crisis. In this set of factors, the success of new left populist discourse blended precarity, re-valorisation of the local state, and blocking the far right. It was only in 2015 that the election of a left-wing, green, social movement coalition started questioning and then transforming the 'smart city' vision to one that is much more citizen-centric and grounded in notions of citizenship and right to the city (Ribera-Fumaz, 2019, p. 201).

In the first year after taking office, the new city administration froze the issuing of new contracts for 'smart city' initiatives and undertook an evaluation of existing ones, publishing 'Barcelona Ciutat Digital: A Roadmap Towards Technological Sovereignty'.[9] Here, Barcelona as a 'smart city' is re-envisioned as an "open, fair, circular and democratic city and a referent in technological policy for a clear public and citizen's leadership"; 'Technological Sovereignty' refers to the notion that technology should be orientated to and serve local residents, and be owned as a commons, instead of applying a universal, market-orientated, proprietary technology: an ideal that seeks to reframe and use 'smart city' technologies to enact radical democracy, where technology serves to empower citizens, to protect their privacy from abuses by the public and private powers, to fight against corruption and to advance towards a more equitable and sustainable economy – according to Gerard Pisarello, First Deputy Mayor of the city council and strong man in Barcelona in Comú (in Galdon, 2017).

Rather than designing the technological infrastructure first and then figuring out how best to use it, Barcelona is applying existing technologies to solving everyday problems like pollution, affordable housing, and congestions. For instance, the city took a bold approach to congestion by closing down parts of the grid-shaped town (Superblocks or *superilles*) and by introducing 300km of new cycling lanes, rather than relying solely on congestion-avoiding apps and sensors.[10] Equally, with regards to data, the city's policy is prescriptive, following the principle that data should belong to the citizens themselves. By opening up data sets in a secure way, Barcelona aims to stimulate local businesses and civic initiatives: "We regard data as a utility like water, electricity or roads," says Francesca Bria, former commissioner of Technology and Digital Innovation of the city (see Morozov & Bria, 2018). Further, service provision (electricity and water) has been municipalised and there are experiments with universal basic income and forms of rent control (e.g., the development of an on-line map and register of vacant properties and forms of sustainable tourism through a FairBnB pilot, which aims to increase the supply of affordable housing). Arguably, one of the most interesting projects developed during these years is the platform Decidim.barcelona, where the city council works alongside techno-political activists coming from the 15M-*indignados* movement, becoming one of the technological cornerstones of the municipal government's democracy programme.

With the new election results in 2019, Colau remained mayor of Barcelona but in a coalition with the Socialist Party. So far we can say that 'technological sovereignty' as digital strategy has disappeared and been replaced by a new policy direction called 'technological humanism' (Cardullo & Ribera-Fumaz, forthcoming). According to Bonet, visible figure of Barcelona PSC, "technology can improve our life, but only if we prioritise people over the few," thus proposing a 'third way' in the digital city policy: to make Barcelona the global capital of technological humanism.[11] On the one hand, technological humanism reasserts the need for new regulations, ethical debates in a citizen-centric 'smart city', and democratic data governance. On the other hand, however, the new policy direction reinstates traditional themes of the neoliberal 'smart city' trope: international competition in the markets of Artificial Intelligence and Robotics, attraction of foreign investments, and governance at a distance of citizens.

The move towards 'technological humanism' seems thus inspired by, or in line with, the neoliberal mantra of Barcelona 'Growth Machine': that the digital revolution needs to be projected as an opportunity for progress and Barcelona, thanks to its traditional eclectic and diverse digital and cultural ecosystem, "could lead a city project that seeks a global visibility similar to that obtained with the 1992 Olympic Games."[12] The new discourse pivots around well-established political figures and the epistemic coalition of exclusive economic circles[13] and academic centres of the region – for instance, around Jose Maria LaSalle, former Junior Ministry of Culture and important voice on the liberal side of PP. LaSalle is shaping this discourse, for instance organising the debate from the launchpad of Circulo de Economía (the liberal lobby of Barcelona and sponsored by Telefonica). Parallel to this academic interest, the 'growth machine' newspaper for Barcelona, *La Vanguardia*, reinstates the aim at attracting foreign capital while investing in techno-ethics linked to Robotics and AI, for instance with opinion articles from its deputy director Miquel Molina[14] (Cardullo & Ribera-Fumaz, forthcoming).

Either way, Barcelona 'smart city' policy seems to be back on the move: 'reloaded' into the city goal of becoming a leading forum on the global scene. The process has presented continuities and ruptures with the past (Ribera-Fumaz, 2019), and these can be grouped under the policy trajectory of 'technological sovereignty', first, and the 'technological humanism', later. The capital of Catalonia has moved, in fact, in between being a key player for 'alternative' techno-politics and digital democracy experiments, vehiculated through its techno-aware social movements, civil society, and city administration, and a leadership role in the 'wordling' discourse, with the promotion of high-profile events and city coalitions (Charnock et al., 2019; March & Ribera-Fumaz, 2016). The real objective of this new policy trajectory is the creation of Barcelona as a great technological hub: the 'smart city' reloaded or technological capitalism with a human face (Cardullo & Ribera-Fumaz, forthcoming).

This is the paradox that Barcelona 'smart city' policy is facing: on the one hand, the 'wordling' process of leading, or actively participating in, a global agenda shaped by transnational digital capitalism; on the other hand, a push towards localism and grassroot initiatives, born within a tradition of cooperation and community organising and more recent anti-austerity movement. Barcelona, in fact, can count on a strong and highly active civil society that rides the wave of the anti-austerity Indignados Movement (also known as 15-M) that began in 2011. The city has indeed a long history of urban movements and strong neighbourhood self-organisation, dating back to the First Spanish Republic (Nel·lo, 2015). Moreover, Catalonia and Barcelona in particular have strong traditions of citizen co-production and direct democracy, with several referendums and participatory budgeting experiences (Della Porta, Fernández, Kouki, & Mosca, 2017). The anti-austerity protests re-activated urban movements. One of the most active current movements is the anti-eviction movement (PAH), co-led by the current mayor of Barcelona, Ada Colau, which aims at stopping banks from evicting people from their houses if their mortgage repayments fail. More recently, secessionist mobilisations for an independent Catalonia from the rest of Spain seem to have activated civil society further.

Either way, Barcelona has sought to *re-politicise* the 'smart city' and to shift its creation and control away from private interests and the state toward grassroots, civic movements and social innovation (March & Ribera-Fumaz, 2016). According to Kitchin (2019a), Barcelona government is asking itself and its citizens "challenging questions: What values do we want to pursue? What kind of society do we want to create and live in? And even, do we want to become a smart city at all? As such, it is practising little, if any, ethics-washing."[15] Following the case study of Barcelona, the rise of citizens' awareness and mobilization seem not a sufficient condition for institutional change. This had to be incorporated in the political agenda through political party action to be able to produce change in the governance model. Conversely, sustained citizens' and social movement pressure led to a paradigm shift from the business – politics collusion 'model' of the 2011–2015 administration (Bakıcı, Almirall, & Wareham, 2013) to the citizen-centred 'model' in which the political leadership and mobilised civil society together implement the alternative 'smart city' vision. For its relentless efforts to include, dialogue, and promote social movements and the commons, I would give Barcelona the leading title of "the democratic city."

Concluding remarks

The neoliberal drive through which the 'smart city' has been implemented steers city services towards marketisation, as these services are further privatised for the benefit of corporate high-tech. Active forms of marketisation have thus been taking place throughout EU and elsewhere in the world in the name of the 'smart city' adoption agenda, but: Who is leading such a process? And, is it reversible? In other words, can cities or groups of citizens say "NO" to such a pervasive technological change and steer the policy debate towards social goals?

The examples seen thus far are a series of measures cities and advocacy groups have started to put in place aiming to safeguard rights and entitlements of their citizens, as well as to act as first player in the data bonanza. Therefore, the notion of urbanism that emerges from this discussion is that of an incipient municipalism: cities that are taking technology and planning control back into their hands, from unhealthy federal policies (such as the United States' very controversial Net Neutrality regulation) and preying corporations (for instance, Uber and Airbnb's 'disruptive' service provision). While the (neo)liberal take on ethics and adoption of the 'smart city' agenda suggests never ending enabling conditions (questions: 'When?' and 'How to?'), an *alternative* version of smartness these cities put forth started asking the challenging question: 'What if?' This question is inherently political and implies broad discussions and the implementation of a normative framework based on human, social, and liberal rights; it also entails the systematic deployment of ethical solutions, such as open source, open standards, and sustainable and ethical procurement; finally, an *alternative* understanding of the 'smart city' reverts practices of poor accountability, prioritising ethics committees and regulations and political oversight.

First of all, *alternative* to what? *In primis*, to the entrenched market principles of efficiency and city growth, that is the neoliberal logic that boosts the current 'smart city' discourse. Second, *alternative* to the understanding that technology is 'neutral', since this is always a composite assemblage of sociotechnical, ethical, and political choices: therefore, that technology does not *per se* solve the problems of humanity and its effects are not the same, or necessarily fair, across the spectrum. Finally, an *alternative* or 'intelligent city' would put people's problems first and then adapt existing technology *to these ends*, rather than starting from the problem *of adoption* of new technology. As this chapter has suggested, effective governance and policy rather than technology *per se* are the drivers for meaningful change with regards to urban issues and social problems. This theme will be explored further in the next chapters which draw from a communitarian perspective, working through the author's direct observation and his research on secondary sources, in relation to a shared wireless network in London and the provision of a public Internet and data infrastructure.

Notes

1 www.aaronswartzday.org/danah-boyd-the-great-reckoning/
2 *https://tada.city/en*
3 https://tada.city/en/nieuws/tada-and-the-city-of-amsterdam-the-first-six-months/
4 https://tada.city/en/nieuws/amsterdams-city-director-explicitly-state-what-youre-doing-with-data-and-why/
5 www.amsterdam.nl/bestuur-organisatie/meedenken-meepraten/openstad-online/
6 www.nytimes.com/2016/12/07/technology/new-orleans-airbnb-model.html
7 www.homelessdublin.ie/content/files/DRHE_November_Homeless_Infographic.pdf
8 www.irishtimes.com/news/social-affairs/number-of-homeless-people-climbs-above-10-000-for-first-time-1.3840762
9 https://bcnroc.ajuntament.barcelona.cat/jspui/bitstream/11703/99399/1/BCN_Digitaleng.pdf
10 www.vox.com/energy-and-environment/2019/4/9/18300797/barcelona-spain-superblocks-urban-plan
11 https://elpais.com/ccaa/2019/
12 www.lavanguardia.com/opinion/20191012/47909438856/barcelona-human-centered-city.html
13 For instance, Cercle d'Economia https://cercledeconomia.com/es/el-cercle/ & The Open Society Foundations, founded by George Soros, www.opensocietyfoundations.org/who-we-are
14 www.lavanguardia.com/cultura/20200112/472817423871/humanismo-tecnologico-barcelona-mobile-robotica-inteligencia-artificial-computacion-cuantica.html & www.lavanguardia.com/cultura/20200209/473368570054/capitalidad-barcelona-madrid-sanchez-colau-barcelona-global-cultura-ciencia.html
15 www.rte.ie/brainstorm/2019/0425/1045602-the-ethics-of-smart-cities/

7 Towards a public service Internet?

In the previous chapter, I discussed various strategies that cities around the world started to put in place in order to respond to the side effects of platform capitalism, digital and networked technologies, and algorithm-led processes of urban governance. At times, cities' interventions are more than just a response to an established conjuncture: some cities, like Barcelona, started to establish alternative strategies able to imagine urban futures that do not necessarily depend on the tropes of 'smart city' as seen thus far; instead, trying to steer the debate on sustainability, equality of opportunities, and fairness in the allocation of public resources and services.

This chapter moves the debate forward going to the roots of the 'smart city' dream-world by way of discussing the current changing morphology of the Internet provision between technological ideals of low latency and hyper speed and the libertarian ethos through which the Internet has been imagined thus far. By showing a variety of possibilities in the 'last mile' Internet provision and highlighting their long-term sustainability issues, it frames this as a critical infrastructure that can be delivered by municipalities. The next and concluding chapter will extend the debate through the author's direct involvement and action research around a community wireless network and ethical hackers, debating commoning and maintenance of the commons in the 'smart city'.

Although 'smart cities' are not just about broadband connectivity, access to unlimited, super-fast and possibly free Internet has been symbolically crucial to their making: the Internet is thought to be the 'smart city' backbone, "the backbone of modern society, a platform for businesses, governments and citizens to exchange news and views, as well as to provide services, whether essential or trivial" (The Digital Agenda for Europe).[1] In the scholarly community, the Internet backbone is often related to research papers from an engineering background, which discuss network protocols, resilience, and security, but not social ecologies and political issues around its deployment. Only more recently, media studies have gone through the so-called 'infrastructural turn' with the aim of focusing on "scale, industry logics, and policy" in the everyday uses of digital platforms, on the "various material assemblages" that make them, and on "the business of Internet service provision" such platforms attend to (Plantin & Punathambekar, 2018). It seems to me that the current scholarly debate swings between a sort of revanchism around the "broken Internet" in the era of digital capitalism (e.g., Zuboff, 2019), and the liberal rhetoric of the Internet as a human right.

Pioneering geography (Dodge & Kitchin, 2003; Graham & Marvin, 2001), sociology (Castells, 1996), and media studies (van Dijk, 2006), and more recent cultural studies research (Hu, 2015; Mattern, 2019) provide an excellent entry point in the debate. As seen in Chapter 1, the tropes on which the 'smart city' discourse is based – consisting of fully automated processes, ubiquitous and seamless connectivity, low latency and real-time responses, and 'everyware and everywhere' (Greenfield, 2013) – are not new imaginaries: they resurface from the 'digital society', the 'informational society', and the 'Internet society' debates. The cultural construction supporting such tropes is sometimes bordering utopian science-fiction, presenting an underlying optimism that emphasises opportunities rather than the risks related to technological adoption. In the middle 1990s, the network society was imagined to organise its relationships around medial networks and nodes that were replacing social networks and face-to-face communication altogether (e.g., Castells, 1996). During the early 2000s, when Wi-Fi protocols were standardised, Bar and Galperin suggested that "it is possible to imagine a future in which ad-hoc networks *spontaneously* emerge when enough Wi-Fi devices are present within an area" (Bar & Galperin, 2004, p. 274).

As Hu suggests (2015, p. 97), "the cloud" is a catchy metaphor for the Internet: "a cultural fantasy, always more than its present-day technological manifestation." It is true that a new generation of wireless network technology has been introduced every 10 years, with some notable advances: "The second generation, aka 2G, allowed for voice transmission, 3G ushered in the app revolution, and 4G brought a drastic speed boost." (Mattern, 2019) While waiting for the next wave of technological transformation (6G, probably), cities are invited to adopt and transform their urban infrastructure to 5G connectivity. This promises to deliver speeds 100 times faster than that offered by previous-generation tech, as well as a network that can support 100 times as many connected devices, summarises Mattern (2019); additionally, "latency, or lag-time, will be cut from roughly 50 milliseconds to 1." I would argue that 'next-generation' thinking – the idea of progressive and dramatic technological change for the better – has been crucial to the development of the Internet and to the symbolic construction of 'smart city' futures. The vast majority of research on 5G technology addresses the challenges and opportunities to push through the 'next generation' wireless network technology. However, few of these accounts question the implementation of 5G systems as such, instead perceiving obstacles as technological challenges to overcome: a never ending series of enabling conditions leading to depoliticised and technocratic responses. Following Dodge and Kitchin 's (2003, p. 6) suggestion that "technological development is grounded in social, political and institutional geographies and discourses that need to be carefully deconstructed", the chapter considers an array of sociotechnological issues that go beyond the variable of speed or latency, such as social needs, policy goals, cultural divides, and forms of governance of the data and service provision, all feeding into the evolving ecologies of implementation, control, and maintenance of the Internet as an infrastructure. It argues for a public-owned provision as a great policy goal for municipalities: ideally, a basic service for the many, a critical data infrastructure for the economy, and a more regulated media-communication network for the *polis*. The Internet provision and governance I think of is an infrastructure for citizens and the city (Internet of People), rather than for consumers and corporations (Internet of Things).

It appears evident to me the new trajectory the Internet as connectivity network has been taking in the 'smart city' discourse and practice: this has moved from a democratic and libertarian ideal, which aimed to favour social inclusion, learning, and seamless communication among users, to a critical data infrastructure which connects algorithms exploiting data pools from consumers and devices – in a nutshell, from the Internet of People (IoP) to the Internet of Things (IoT), from a network of social relations (e.g., the Network Society) to islands of connectivity for 'things' (e.g., the World Wide Herd). What implications are these changes presenting for the issues of democratic governance, citizen participation, and digital rights this book is concerned with? What future scenarios are the 'smart city' proponents fostering for the lay person and the 'citizen'? How can cities develop an infrastructure on which they can exercise more control while being *useful* for their citizens? While the chapter will not provide definitive answers to such questions, it starts debating these critical points.

The possibility of cities and communities running a public service Internet has only received little public attention thus far, but is of huge importance in the context of how we can best overcome high-speed superficial online communication; distribution monopolies (such as AT&T or Verizon); digital monopolies (Google/ YouTube, Facebook, or Amazon); and an array of ethical issues concerning privacy, commercial tracking, and data-ownership. As Mazzucato (e.g., 2011) has been suggesting for some time, the private sector and the dominant neoliberal discourse downplay massively the role of the public in driving innovation and growth via core infrastructures, investments in skills and literacy, and more generally in taking on board the political-economic risks of failure by socialising the costs. Although fast connectivity is crucial to the development of data capitalism, the 'backbone' on which networked technologies and people exchange data is often assumed. Both private and communitarian actors often gloss over the fact that the Internet as infrastructure is a form of 'public value' (Mazzucato, 2018a) embedded in a longer-term strategy of technological and infrastructural innovation. Moreover, as Hu (2015, p. xvii) reminds, "the all-but-forgotten infrastructures that undergird the cloud's physical origins [are] often originated in a state's military apparatus."

What alternatives to the commercial Internet are desirable? How can a Public Service Internet be created and financed? What is the role of citizens in the governance of this infrastructure? The two case studies in this chapter, and a third one in the following chapter, suggest different ecologies of implementation, control, and maintenance of the network which delivers the Internet in cities, thinking through *how* and in *whose* terms this infrastructure is developed in cities, and by the cities themselves.

Becoming a 'smart' Thing

Kitchin's table (2016a) reproduced in the Introduction of this book shows the diversity of 'smart city' initiatives and 'smart' technology applications, ordered through their different domains of deployment (Government, Security and Emergency Services, Transport, Energy, Waste, Environment, Buildings, and Homes). To this variety of applications corresponds the multiple ways in which the 'smart city' is implemented: "smart city interventions are always the outcomes of existing social and spatial constellations of urban governance and the built environment" (Shelton et al., 2015, p. 14). The ways in which 'smart city' policies travel across places widen the gap between the 'smart city' idea(l) and "the actually existing cities, territories and relationalities where these policies are being constructed and implemented" (p. 22). This variety of 'smart' technologies, projects, and initiatives presents, however, some rather consistent characteristics. For the purpose of this chapter, I try to group them in the following three points.

First, smart technologies are always somehow linked to an algorithm-led response, even only for a card payment or for an instance of communication. Ultimately, algorithmic functions represent the 'smart' bit in smart technology. They *always* involve some transmission of information or data from one agent (machinic, human-operated, or human) to another. Recording, storing or transmission are the *modus operandi* of smart technologies – their reasons to be. Therefore, while traditionally the Internet has facilitated interaction between humans, the current sociotechnological landscape starts being dominated by different configurations we can group under the Internet of Things (IoT). These 'things' are digital devices that typically have a communication interface, processing and storage units, and sensors for detection of environmental changes or for service provision to other clients. The combination of short-range mesh networks and the wider cellular network can provide wireless connectivity to these 'things' in order to exchange data to the wider Internet. It is data, not communication between people, that really matters to high-tech industry today. Data power the new horizons of connectivity and future imaginaries of technocracy, from the 'Industry 4.0' to the 'smart city'. While connections between people favours a degree of communication that would be otherwise impossible (typically in Voice Over IP calls across the globe), algorithms that regulate many other aspects of daily life, such as financial interfaces and banking transactions or IoT devices at home and in cities, exchange information in milliseconds in mostly automated and sometimes autonomous processes. A different kind of "experience" is now being promoted, for instance a user dealing a holographic communication system for a "tangible, immersive, and interactive communication experience" (Li, Clemm, Chunduri, Dong, & Makhijani, 2018).

Second, it appears evident that this variety of smart technologies are in desperate need of some form of connectivity or infrastructural support for data exchange. Heterogeneous 'smart city' objects connect via networks that use short-range communication technologies and, eventually, Internet gateways. In the future 'smart city', providing flawless connectivity will probably become a real challenge as the density of connected devices that have multi-radio capabilities increases. To this regard, 5G giant Huawei suggests to evolve connectivity via a new "extra dynamic IP addressing system" (Li et al., 2018), while Dell proposes the development of the "World Wide Herd"[2]: the two Big Tech aim to create islands of hyper-connectivity which permit gadgets, devices and nodes inside the mesh to communicate instantly with one another without having to ship data throughout the web – for instance, driverless cars could roam through islands of data exchange within the proximity of the 'herd' of data-led devices. On the one hand, the battle for speed and reliability of Internet connectivity has become synonymous with city growth, competitiveness, and progress: in other words, a brand, like in the case study below of Chattanooga city. On the other hand, inequality of access (for people and 'things') can be projected as a paradox at the heart of the smart city: how can this super connected and increasingly complex urban environment work without a strong backbone and its capillary distribution to as many users/customers/citizens/'things' as possible? The case study of New York City below proposes a different possibility with regards to the Internet of People (IoP) for the many.

Consequently, and third, a major obstacle for the adoption of smart technologies seems to be exactly this scarcity, which capitalism is so good at creating via market imperatives of privatization and efficiency: Fibre-to-Home business in the United States, for instance, is in the hands of Big Cable which hold quasi-monopolist positions in the US spectrum, for historical reasons (Dodge & Kitchin, 2003; Hu, 2015). These giant providers control Internet provision, acting as gateway for people and 'things', metering and throttling connectivity at a high price and slowing down at will . In the fast-growing 5G business, the same accumulation patterns are already visible: the US government is said to be holding "a big auction of the wireless spectrum used for the [5G] technology. The spectrum could be sold to companies like Verizon and AT&T" (D. Clark, 2018). Would the proposed 5G connectivity rather increase the exclusivity gap (super connected areas or hubs) and splinter service provision further (Graham & Marvin, 2001)?

These three points support the argument for the changing morphology and the growing importance of network connectivity and data infrastructure in the 'smart city' – and the coronavirus crisis massively underscored this, as I will argue in the final chapter. These might be planned by cities as a public service for the many or as an infrastructure delivered by the market for the few. Cities and communities alike are already paying for the 'scarcity' through which the Internet and its related data infrastructure are delivered aided by the market's invisible hand: with higher rates and artificial caps, poor rights, ethics and governance records, and the splintering of the service. Moreover, the variety and pervasiveness of smart technologies in the everyday life of cities raise a formidable array of ethical issues in relation to extraction and treatment of data, surveillance, and control, as well as privacy, consent, and disclosure of personal and sometimes sensitive data for any purpose. Morozov calls it "data extractivism" (2013; see also Mezzadra & Neilson, 2017), others "data colonialism" (Thatcher, O'Sullivan, & Mahmoudi, 2016) or "platform capitalism" (Shaw & Graham, 2017b). They all point to the extraordinary concentration of data produced through uses, consumption, and leisure – that is, through life itself – in the hands of a few Silicon Valley start-ups. Participation through the workings of smart technologies and the Internet can thus come to the price of reduced privacy, nudging advice, and more severe breaches of collective rights, such as mass surveillance. Therefore, the chapter advocates for the regulatory role of the public, as both municipal service and democratic form of governance, over the 'last mile' provision of the Internet and the overall organisation of the infrastructure which delivers 'data' (access points, standards, data ethics, maintenance, data centres, etc.).

By developing, maintaining, or controlling their own Internet grid, cities should be able to negotiate more favourable and ethically binding 'last mile' access for their citizens, for instance by leveraging their contractual weight towards digital providers and platforms. A public Internet grid would allow municipalities, or their alliances, to have more negotiating power towards digital platforms – for instance, when setting new urban policies, standards, and regulations for companies such as Airbnb or Uber – and to foster ethical procurement towards tech providers. Social movements and interest groups can set standards of data ownership and ethics of data extraction, treatment, and reuse – perhaps setting their own TADA manifesto, audits, and feedback loops. In an 'intelligent city', the governing bodies and social movements would own and deliver their best data infrastructure for the common good, based on rights and a municipally owned Internet service.

This is also a matter of providing a *useful* utility for all. I consider the Internet, in fact, a critical urban infrastructure, like electricity or transport; one right that makes the "right to the smart city" operative (Cardullo, 2019). In the following sections, I present two examples of Internet infrastructure delivered by municipalities. These are very different in the aims and the forms of governance they present. They are two case studies I have been following for some time, mostly based on secondary sources. The next chapter presents, instead, a case study based on primary fieldwork and direct action research in the field: together, they advance the argument of a publicly owned Internet and data infrastructure for the common good.

Chattanooga, the "Gig City"

Chattanooga, Tennessee, was the first city in the United States to offer 1 Gbps high-speed Internet (at the time, over 200 times faster than the national average) in 2010. This mid-sized Tennessee city (170,000 people) made the headlines again in 2015 when it implemented the world's first community-wide 10Gbps Internet service. The city transformed its image of a polluted and failing one into the thriving "Gig City" (Kitheka, Baldwin, White, & Harding, 2016). More importantly, this super-fast infrastructure has been offered via the municipal and non-profit Electric Power Board, making it the largest public investment in the United States on the matter (EPB, 2015).

This is particularly relevant in the context of the United States, where Internet activists have long complained of the widening digital divide caused by private contractors dis-investment (the so-called Big Cable: private giant providers such as Comcast and AT&T), which have left behind small and rural towns and impoverished neighbourhoods (Gonzalez, 2018). Moreover, the EPB fiber in Chattanooga disrupts powerfully the Net Neutrality debate – that certain types of data should be throttled (slowed down), connections metered or charged at different rates, and certain sites blocked. At the present, these are hotly debated topics in the United States as a consequence of the recent coronavirus crisis, with the generalised 'lockdown' of population, and the repealing of neutrality rules by the Federal Communications Commission (FCC): to this regard, the cities of San Francisco, Seattle, and Boston are promising municipal broadband as an alternative which seems to gain momentum, especially among young voters.[3]

The Chattanooga case study offers interesting points for the debate on public infrastructure, governance, and commoning in the 'smart city'. First, it dismantles a few myths: that the public is always behind the technological curve, and thus needs private consultancy, interventions, and skills (see Kitchin, Coletta, Evans, et al., 2017; Mazzucato, 2018a); and that the public is slow, unreliable, and inefficient in delivering – EPB completed its ambitious project much earlier than planned using a federal loan issued to implement cables for its electric smart grid (Davidson & Santorelli, 2015). Just to give a sense of this town's achievement, in September 2016 the European Commission adopted its strategy on connectivity, a European "Gigabit Society," with the main strategic objective to give 1Gbps to all schools, transport hubs, and main providers of public services and digitally intensive enterprise by 2025: Chattanooga city realised this goal, at a much smaller scale, 15 years earlier (in 2010) and did it as a developmental policy funded by public investments (that is, not solely to overcome "market failure", but as a strategic and planned political decision).

Second, the case study highlights the relevance of the municipal scale vis-a-vis county, state, and federal scales, whose legislation have hindered rather than facilitated it (e.g., positing limits to EPB pricing policy and geographical expansion). In other words, the city inverted the cycle of privatizations and market-led solutions in a particularly hostile cultural and normative environment: in order to guarantee the 'free market', Big Cable and the Conservatives brought forward many legal challenges to Chattanooga city, and even TV commercials warning against the perils of public investments (Rushe, 2014). Similarly, Comcast produced an advertisement for almost one million dollars in a campaign against the municipalisation of the Internet provision in Fort Collins, Colorado, where a public consultation in 2017 was overwhelmingly in favour to take the Internet back into public hands.[4] Almost every municipal and community-owned broadband services in the United States have some background stories on how they have been struggling with the changing legislation at the different scales of political governance.[5] According to a report by Next Century Cities, "over half of the country's estimated unconnected population – who do not have access to broadband – live in states where municipal networks are barred or outright banned by state legislation."[6] Currently, there are 19 states that limit local authorities to build networks or partner with local companies,[7] although BroadbandNow's report has identified 25 states with laws on the books that are designed to impede municipal broadband initiatives.[8]

There is little doubt for me that Chattanooga has a typical 'smart city' strategy in mind: city growth and attraction of a creative class via digital hubs, which now bolster several tech incubators and attract new businesses, including venture capital funds (Rushe, 2014). Although the material and symbolic effects of this 'Internet boomtown' (Koebler, 2016) appear solid on the local economy (Malmgren, 2017), these have also been assumed to trickle down to the poor. A degree of scepticism that super-fast Internet translates automatically into greater benefits for lay people is, however, due: if it takes "just 33 seconds to download a two-hour, high-definition film in Chattanooga" (Koebler, 2016), then we might want to ask how many films, or similar content, an average family can possibly use in one day? As an indirect evidence, EPB has now nearly 100k Internet customers (more than half of the civil and commercial residents of Chattanooga), which drops to about 20% in its poorer neighbourhoods (Koebler, 2016): despite EPB offering half-price subsidised Internet for families that have students enrolled in school lunch programs (Netbridge), it appears that poorer residents prefer signing up with Comcast which offers a cheaper service although much slower and capped. In other words, what is the real *use value* of super-sized 'smart city' initiatives like this?

Moreover, issues of governance are case-specific and they require deeper ethnographic engagement and more longitudinal research to fully evaluate the impact of 'the Gig' on the overall population, a third of which is said to be at risk of poverty and social exclusion. It is reasonable to believe, however, that citizens can lobby the mid-sized town administrators more successfully than when acting as individual customers dealing with off-shore call centres funded by Big Cable. A public-led provision of the Internet as I envision, in fact, would not work without different degrees of involvement from users and communities of practice. I believe there is plenty of intelligence in the ethos and practices of community-led projects and in their informal uses of infrastructures, which cities can harvest in order to become more technologically inclusive and competitive.

NYC: between inclusion and exclusivity

The city of New York was announced as the Best Smart City of 2016 at the Smart City Expo World Congress in Barcelona on four pillars: LinkNYC (7,000 planned high-speed multi-functional kiosks); Marketplace.nyc and Urban Tech NYC; a programme for piloting and scaling smart technologies; and a set of comprehensive guidelines for ensuring the equitable deployment of 'smart city' technologies. While the last objective remains a politically engaging principle difficult to pin down and requires further research, we know a good deal about the first from the triangulation of many sources.

On the one hand, the process of rolling out the advertisement-funded and Alphabet-owned LinkNYC units seems to have been widely accepted with enthusiasm – although some critiques appeared in traditional and social media with regards to the geography of their implementation (e.g., priority given to the super-gentrifying areas of Manhattan) and the uses that some fellow citizens have put in place (e.g., complaints towards homeless and young people's "occupation" of sidewalks, with consequent curtailing of unrestricted browsing).

On the other hand, the reality many NYC residents face is staggeringly unequal with 46% of families living below the poverty line who don't have service, usually due to high prices. It's worse in minority neighbourhoods, long neglected when it comes to broadband upgrades: "18% of residents – more than 1.5 million New Yorkers – have neither a mobile connection nor a home broadband connection."[9]

Overall, two-thirds of New Yorkers are renters in a federal state that still holds on to some rent-stabilizing policy – and many housing activists are now making links between inequality and the hype around the 'smart city'.[10] Using our geographical imagination and the availability of digital information, we can start plotting some patterns. If our digital *flaneur* took an augmented walk through Brooklyn South on Park Slopes, she can stop outside a housing block built in 1905: this has one of the new LinkNYC stations outside. She can charge her mobile device there using a super-fast connector while browsing the Displacement Alert Project (DAP) map. She would then see the building in front of her being colour-coded as at extreme risk of displacement.[11] The dashboard-like map is made by the Association for Neighborhood and Housing Development (ANHD) and this map shows as high as 26% of the buildings in NYC have a high-risk score. This score is given by a Combined Building Risk Indicator which takes into account the recent *per-unit sale price of the building* when this is greater than an average neighbourhood price, *NYC Department of Buildings work permits* (they indicate apartments that have become vacant and been upgraded by the landlord, often allowing for major rent increases), and *the percentage of rent-regulated units lost in each building*.

ANHD, like other counter-mapping initiatives such as the Anti-Eviction Mapping Project (AEMP),[12] produces also an "affordable housing risk chart". This shows a significant increase in the 'Percent with Severe Overcrowding' across many neighbourhoods, and for the first time in Brooklyn: "Severe overcrowding is the underwater portion of the homelessness iceberg," they suggest. An increase in overcrowding captures families that are doubling-up in order to resist gentrification pressure of displacement, therefore claiming their "right to stay put" in the neighbourhood (Newman & Wyly, 2006). Brooklyn, with over 300,000 rent-stabilized apartments, the most in any borough, is emblematic of New York's housing emergency. It is the largest concentration of African-Americans in the United States and it has been attracting hyper-investment in its real estate since 2011, led by more affluent buyers, denounces ANHD. Our digital powered ride around Brooklyn repositions the 'right to the city' as "the issue of the direct access of the means of existence, production and communication" (De Angelis, 2001), a right to inhabitation for a start. 'Rent' and utilities are in fact to be considered as material practices which determine the very existence and perpetration of the city as a commons (see Bresnihan & Byrne, 2015).

The important point about counter-mapping is that these associations make a visible link between the regeneration wave of technology adoption and the worsening housing crisis in their neighbourhoods, between the city's real estate and the techno-utopian growth machine. And in order to do so, they use data sets and mapping techniques proper of the 'smart city' dashboarding tradition and visuality (Stehle & Kitchin, 2020), with almost real-time inputs, GIS manipulations, time-series indicator data, interactive maps, and more qualitative indicators such as digital storytelling (da Silveira Arruda & Yances, 2016).

"But we will take matters into our own hands," NY Mayor De Blasio famously tweeted after Trump exited the Paris agreement on climate change, in May 2017. The tweet was perhaps symptomatic of a change of policy at the heart of the Big Apple. The same year, a small percentage of NYC's budget, $10 million supported also by a federal grant, was devolved to a pilot for the provision of a free 25Mb Broadband service to Queensbridge Houses, Long Island City. The achievement of this pilot, now extended through April 2020, is not so relevant compared to mainstream talks of hyper connectivity, speed, and resilience of the network. The legal definition of broadband in the United States is in fact 25Mb download speed or more, at the present.[13] What makes 'Queensbridge Connected' noteworthy is the sheer number of people involved: Queensbridge is the largest public housing estate in the United States with over 7,000 low-income residents spread in 95 buildings; that makes over 3,000 residential units in one neighbourhood, a small town indeed.

The city, acting as a big and authoritative buyer for several thousand people, managed to negotiate a reduced price for the piping of the cables (Fibre to Home) and for the supply of Tier 1 broadband, thus making huge savings to devolve to a socially just cause: to provide basic free broadband to the many, so that "students will have an easier time doing homework, parents can look for jobs and access resources," as the neoliberal rhetoric of the self-reliant individual demands. Of course, more heavy users can purchase a premium rate broadband at very competitive prices: no curtailing of individuals' freedom or their entrepreneurial spirit, either!

The vast network of hotspot routers was given via an ethical tender to Spot On Networks, specialist in multi-residential connectivity. This move followed talks with the residents, who signed a consent form even if, technically, a consent was not required. In addition, some local residents have been employed in order to ask for consent and provide information to the residents (who were mostly concerned about their own privacy); for small maintenance of the network (no cold call to overseas call-centres); and for outreach of local projects (a digital radio, a local bulletin, etc.) (Lewis-Kraus, 2016).

NYC presents itself as an example of cities becoming more self-reliant from federal politics while teaming up with other municipalities in order to share their own experiences; a city which seems to be led on values and principles of social justice as with regard to their stance on technology. NYC "Broadband for All" is a state-level political project where high-speed connectivity will be provided to all who need it by 2025[14] – in a city where currently 22% of households do not have access to the Internet, and in a country where, according to a report commissioned by the Roosevelt Institute, the average household spends more per year on communications services, most of that on wireless and broadband, than it does on electricity or gas or grocery (Mabud & Seitz-Brown, 2017). Recently, the Mayor's Office of the Chief Technology Officer released '[The] Internet Master Plan',[15] a new proposal which aims to create a citywide open access broadband network, bringing faster and cheaper service to citizens who desire to subscribe to it. The Internet Master Plan identifies the partnerships and infrastructure required with the ambitious goal to end the digital divide in the city. According to the city plan, achieving universal broadband will require lower-cost options for home and mobile service as well as no-cost access at computer centres, in public spaces, and through wireless corridors. New York City will begin sending out a formal request to the private sector for assistance. From there, the city promises to offer its private sector partners broad access to the city infrastructure, including 800 city rooftops and 20,000 poles to deploy both fibre and a neutral radio access network capable of reaching every corner of the city. The city also pledges to bind contractors to the strictest protocols of privacy and net neutrality once completed.

In addition, New York City filed a lawsuit against Verizon, claiming the tech giant failed to comply with a citywide fibre rollout as agreed in its cable franchise agreement and continued its practice of 'digital redlining' poor neighbourhoods.[16] NYC leads US cities' rebellion against the FCC and government decision to roll back measures meant to protect users from privately-owned Internet Service Providers: cities are committing not to tender public project to Internet Service Providers that do not honour Net Neutrality principles. NYC is also the founding member, with Barcelona and Amsterdam, of Cities for Digital Rights, a coalition that wants to protect and uphold human rights on the Internet at the local and global level.[17]

As a consequence of this renewed policy towards connectivity for the many, I would give NYC the very hopeful title of: "the nearly-inclusive city".

Discussion

Searching through successful stories of municipal broadband service in the United States, it appears that most of these initiatives have built their own citywide telecommunications networks based on cabling already operated by a municipal electric utility, for instance the EPB in Chattanooga, Tennessee or the Greenlight for Wilson, North Carolina. Another good example is NextLight in Longmont, Colorado[18]: the service is provided by LPC, the city's community-owned and not-for-profit electric and Internet services provider. This is important because it shows how public services can become cumulative and build upon local strengths and skills, while negotiating provision at a reduced price.

It is important to note that cities in the case studies above have shown a different sensibility and approach to the issue of public Internet connectivity. While New York City has installed a Wi-Fi network through the largest public housing complex in the United States, it does not own the power company, as in the Chattanooga example. While Chattanooga remains a good example of efficient public service delivery, but apparently with limited impact on the marginalised population, New York City experiments in its public housing shows how utilities can be delivered as a public service with redistributive effects, and even foster a more democratic model of urban governance.

A major obstacle for the adoption of smart technologies seems to be exactly the 'scarcity' through which the Internet is delivered, through and against market imperatives of privatisation and efficiency: in the United States, Fibre-to-Home is mostly in the hands of Big Cable (such as Comcast, Verizon, Time Warner Cable, or AT&T): these giant providers act as a gateway for people and 'things', metering and throttling connectivity and slowing it down at will. As a consequence, line-sharing plans at the top level of network provision have been slowly scrapped[19] with increased costs at the end of the pipeline, the 'last mile' service to individual families and companies.[20]

Moreover, it has been the case that Big Cable have used their massive profits to lobby politicians at a different scale, in order to dismiss a 'return to the public': the dominant narrative assumes this being inefficient and cumbersome and claims a breach in 'free market' competition and principles. It is actually striking that giants of communication and cabling systems have put so much effort, money, and political leverage into defending free market and competition, while acquiring monopolistic positions everywhere in the United States: this reminds us that neoliberalism is much less about laissez-faire policy than control of the market for the advantage of private profit. As Foucault suggests (2008, p. 132), "neoliberalism should not be identified with laissez-faire, but rather with permanent vigilance, activity, and intervention." For Peck (2004, p. 394), "markets themselves are not, never have been and cannot be spontaneously occurring and naturally self-regulating." In the prevalent neoliberal framework, public investments in the Internet service are admissible only for market imperatives of economic growth and for supporting areas and communities which have fallen through the net of private providers' profit. The neoliberal strategy for broadband connectivity can be summarized in the catchphrase of "market failure", that is, public investment should only take place where the market *is not* providing the desired connectivity.

To sum up, there is a solid ground to suggest reverting the circle of privatisation of infrastructures and the consequent centralisation of the Internet in the hands of very few and powerful companies. We can observe a growing trend in the US for the consolidation of municipal and cooperative-owned small providers: in January 2019, there were over 800 initiatives, of which 500 are served by some form of municipal network and more than 300 are served by a cooperative.[21] Beyond the neoliberal rhetoric of the "market failure", public resources and skills could be turned towards more fair and open infrastructures, by bolstering different approaches to commoning – the contested process of delivering, protecting, and maintaining infrastructures as a commons. From the Fibre-to-Home service to the remote data farms, in fact, technology presents itself as a composite assemblage where issues of social justice, ethics, and democratic governance are to be addressed, by municipalities and citizens, at each step of the technology stack and carefully evaluated case by case.

The issues of public availability and fair access to the Internet are not going to challenge the many faces of digital capitalism, but they might provide one possible gateway in the life cycle of data for more local control and accountability. Going up on the connectivity stack, a whole set of services, platforms, data centres, clouds, and infrastructure governance would need to be regulated and/ or brought into public hands, whether by the state (for larger infrastructure build, such as the 'backbone') or municipalities (for the 'last mile' delivery and local governance). In the present scenario of a growing demand for connectivity from people to 'things' – and the looming Covid-19 pandemic underscored this point massively – it is easy to believe that the capacity of the networks will be increased and expanded, or capped and slowed down further. We can move the above question to the supply side and ask: *how* and in *whose terms* is this infrastructural development going to happen?

Talks and experimentations around new protocols and technological solutions, such as the up-and-coming rhetoric around 5G connectivity, seem to re-propose a circle of innovation for the few, and for the myriad of 'things' that Big Tech are eager to sell (Autonomous Vehicles, for instance). There is room to believe that the costly deployment of 5G networks will enable digital hubs and super-connected areas of the city (such as Dublin Docklands or New York Hudson Yard) with huge investments which will lock-in cities and urban fabrics for the years to come. Or at least until 6G will present itself as the new frontier of technological progress,[22] a vision for a "better tomorrow – a world in which bandwidth, speed, and growth are virtues in and of themselves" (Mattern, 2019).[23]

Or maybe 6G, unlike its predecessors, won't be about being faster, more ubiquitous, or resilient: "Maybe it will be about energy efficiency. Or local responsiveness. Or slowness. Or reflexivity. Or privacy, or equity, or digital justice," suggests Mattern (2019).[24] From an ethical perspective, the Next Internet will hopefully include the principles of privacy-by-design, avoiding corporate tracking and state surveillance; while from a normative perspective, this will probably include capped slowness (at least 25Mbs, which is currently the limit at which the United States defines 'broadband', updated to 100Mbs by the recent presidential campaign by Bernie Sanders) in order to guarantee the right to be online to as many people as possible. In other words, the public discourse around the 'digital society' needs to be shifted back towards the Internet of People: not smart vehicles without driver or caring robots, but connecting infrastructures and digital services for 'the many'; utilities that become *useful* again for a whole series of purposes that matter everyday practices of working and inhabitation. This strategic infrastructure cannot be left to private and often unregulated, although massively subsided, oligopolies but needs to be steered towards institutions and regulations that work for the common good.

Concluding remarks

Nowadays that the Internet of People (IoP) and Internet of Things (IoT) have merged into buzzwords like the 'smart city' and 'Industry 4.0', access to a reliable and affordable Internet is not just a matter of geographical inequalities, between the city and the rural or between areas of the same city. It is instead a matter of basic participation in civic life, and a means through which the full human being can thrive with dignity: one gateway for the 'right to the smart city'. Whether we like it or not, in fact, digital services are becoming the privileged or only way to pay taxes or parking tickets, to apply for a job or complete school homework, to register a birth or vote: in other words, the way in which people, especially in the Global North, inhabit cities every day (and the pandemic crisis, with the 'stay at home' policy aimed at limiting contagion, has underscored exactly this point).

For some, it is also a matter of taking the Internet (and, thus, data and digital platform economies and the related rights) back into citizens' control and away from the big players and the state. This is a motivation that has traditionally bolstered community networks but it has become, at least in the aims, a policy to be followed by a growing number of cities. Some cities are moving towards the goal of a more universal access to the Internet, like New York; others have taken further steps by fostering data commons, like Barcelona or Amsterdam. Perhaps, not any one city can control the overall process of data production and gain complete 'technological and digital sovereignty'. But, are public funds available to cities for the implementation and maintenance of a public Internet service? And is this a social priority? Indeed, in the currently prevalent neoliberal framework, public investments are admissible only for supporting areas and communities which have fallen through the net of private providers' profit and market imperatives of economic growth. The public strategy for broadband connectivity can be summarised in the catchphrase of "market failure", that is, public investment should only take place "where the market *is not* providing the desired connectivity".[25]

In the electoral plan unveiled recently,[26] the US Democratic Party presidential candidate Bernie Sanders proposes to break with the country's private monopolist tradition of Internet provision aiming "for municipalities and/or states to build publicly owned and democratically controlled, co-operative, or open access broadband networks."[27] The plan echoes the UK Labour 2019 electoral manifesto[28] which proposed the launch of an Internet public provider, integrating the broadband services from British Telecom into a new public entity, British Broadband. Both proposals upgrade the definition of broadband, from the current 25mbps to high-quality fibre of at least 100mbps, for all. More notably, the two plans break with the neoliberal discourse of private provision and weak users' rights (consumers, rather than active citizens), shifting the political debate on the state of the Internet broadband: this is eventually considered a new citizenship right rather than a marketplace commodity, for example by ensuring that all public housing can provide free broadband services to all residents.

Two strategies seem to emerge here: British Labour's proposal considers taking back into state's ownership the provision of the Internet from the ground up as a state monopoly (British Broadband); US Socialists' plan leaves it to cities to organise themselves while applying for available federal public funding, echoing the tradition of American libertarian municipalism (Bookchin, 2017). The following chapter takes forward the issue of municipalism and urban commons, where public provision of the Internet is meant as an entry point on *alternative* or common-oriented 'smart cities'. In the concluding chapter, I ask, among other things, whether the adjective 'smart' is still appropriate, and whether it can be of any use for radical democratic urbanism. Is it rather the case to be going "beyond the smart city today," as the title of a recent conference call in Rotterdam suggested?

Notes

1 https://ec.europa.eu/digital-single-market/en/news/broadband-big-pipes-potential-growth
2 https://blog.dellemc.com/en-us/distributed-analytics-meets-distributed-data-with-a-world-wide-herd/
3 A handful of towns are following now on the example offered by Chattanooga, for instance Hopkinsville, Kentucky, https://hop-electric.com/news/introducing-the-new-energynet/
4 https://arstechnica.com/tech-policy/2017/11/voters-reject-cable-lobby-misinformation-campaign-against-muni-broadband/
5 For instance, some interesting stories here, about Wilson, North Carolina, but the narrative does change much across the board in the United States: https://ilsr.org/wp-content/uploads/2013/01/nc-killing-competition.pdf
6 https://nextcenturycities.org/fact-sheet-the-opportunity-of-municipal-broadband/
7 www.marketplace.org/shows/marketplace-tech/covid-19-pandemic-internet-access/
8 https://broadbandnow.com/report/defining-municipal-broadband-roadblocks/
9 https://tech.cityofnewyork.us/internet-master-plan/
10 https://anhd.org/resources-reports/2017-how-is-affordable-housing-threatened-in-your-neighborhood/
11 http://map.dapmapnyc.org/app/
12 www.worstevictorsnyc.org/about
13 The more recent proposal from the Bernie Sanders' electoral manifesto is to update such a definition to a high-speed broadband of 100Mbps. www.businessinsider.com/bernie-sanders-internet-as-utility-plan-explainer-2019-12
14 https://www.ny.gov/programs/broadband-all
15 https://tech.cityofnewyork.us/internet-master-plan/
16 https://arstechnica.com/tech-policy/2017/03/nyc-sues-verizon-alleges-failure-to-complete-citywide-fiber-rollout/
17 https://citiesfordigitalrights.org/cities
18 https://www.longmontcolorado.gov/departments/departments-e-m/longmont-power-communications/broadband-service/in-the-news
19 www.vice.com/en_us/article/ev8n3e/big-telecom-lobby-says-theres-too-much-broadband-competition-pushes-fcc-to-harm-smaller-isps?mc_cid=4b685d0495&mc_eid=4c4b31ca2e
20 www.newamerica.org/oti/policy-papers/the-cost-of-connectivity-2014/
21 Institute for Local Self-Reliance, https://muninetworks.org/communitymap
22 www.spotonnetworks.com/2019/05/09/what-does-wifi6-5g-mean-for-multifamily-properties/
23 https://reallifemag.com/networked-dream-worlds/
24 https://reallifemag.com/networked-dream-worlds/
25 https://ec.europa.eu/digital-single-market/en/state-aid
26 https://berniesanders.com/issues/high-speed-internet-all/
27 www.businessinsider.com/bernie-sanders-internet-as-utility-plan-explainer-2019-12
28 www.independent.co.uk/news/business/news/free-broadband-labour-plan-internet-wifi-nationalisation-a9205031.html

8 Sociotechnical capital and trust between urban commons and commoning

The two case studies in the previous chapter compared technosocial assemblages of urban governance and sharing in the 'smart city'. The first case evaluated the provision of municipal extra-large broadband in Chattanooga, Tennessee, and addressed the opportunities offered by a successful public ownership of infrastructures, although with a critical eye towards its governance processes. The second case looked at NYC pilot with free broadband to all residents of the largest public housing in the United States. A different role has been played there by the city which acted as local authority in the market for broadband provision and fostered advocates among local residents. The present chapter, instead, focuses on the communitarian project of building a free wireless network in a working-class neighbourhood of London, where I lived for many years during and after my PhD at Goldsmiths (UoL). It unpacks sociotechnological issues at the heart of technology making, wireless connectivity networks, and Living Labs deployment.

The three case studies plot a trajectory along the provision of the Internet as a critical public infrastructure – that is, a basic need for citizens and their participation in the 'smart city'. They highlight a crucial issue at the heart of the 'smart city' yet to come, showing that cities have a lot to learn and gain from the 'public' and the commons: citizens might even be able to shift their role of consumers and regain some privacy and, eventually, control over technologies as the overall framework moves up on the scaffold of citizens' participation (Cardullo & Kitchin, 2019a). As McLaren and Agyeman suggest (2015, p. 1), "good governance and collective city structures" are what successful cities need. Here, of course, the key words are "good" and "collective", which require the articulate response that these examples want to offer. These are ethical and political questions: in what forms these modalities are combined is difficult to forecast and should be evaluated case by case. I contend though that an alliance between city and the commons, rather than the market, can reimagine the adoption of technologies which depend heavily on urban infrastructures such as the Internet of People and Things. As I argued in the previous chapter, the issue of public provision of the Internet is only an entry point on alternative or common-oriented 'smart cities'. Although these are not simply about broadband connectivity, in fact, access to unlimited, super-fast and possibly free Internet has been symbolically crucial to their making.

The case study below revisits an ethnographic account of a shared community asset, a wireless network in inner-city London. Based on primary fieldwork data and years of participatory research with observations, action research, and networking in the field, the chapter takes the reader behind the routing mesh, from cabling and firmware upgrading to social trust and communitarian bonds. It reflects on the commoning practices of its community of interest (an interplay between stewardship and social trust) in a rapidly changing and gentrifying urban space. It opens to a new set of questions: What kind of trust, collective intelligence, and forms of capital circulate between digital infrastructures, citizens, and the city? The chapter makes a positional argument for a 'smart approach' to the commons, advocating for the 'intelligent city' to become a crucial stakeholder in creating and maintaining urban commons. Through notions of sharing, collective intelligence, and social capital, it argues that initiatives concerning techno-politics can foster a high degree of trust and social/political capital, but questions the effective inclusion they enable through two issues: 'stewardship' and 'rent'.

Commons and commoning in the 'smart city'

A useful line of thought in the vast literature on commons suggests that this concerns not so much the conflict between public and private (De Angelis, 2001; Stavrides, 2016). Rather, commons manifests itself, historically, as the *locus* within which peoples' social reproduction is put into practice. The confusion is exactly at the point of departure from the idea of the commons that emerges, in medieval England, as a fight against the enclosures of land. According to Illich (1982), people saw this common space not as a "non-private space", but rather as the space where their everyday subsistence was guaranteed. Using the commons was an act of temporary appropriation, an everyday practice for the purpose of social reproduction: thus, a commons is determined not so much by its proprietary regime, but by the uses and practices around it. A common good, in other words, has no ontological substance in itself. It *becomes* a commons because of the qualitative relationship with one or more subjects; it is the use value, of a place or an object, that makes it relevant to the commons: "you don't have a common good, you share in common good" (Mattei, cited in Iaione, 2012). For McLaren and Agyeman (2015, p. 24), "sharing and cooperation are universal values and behaviours" and, therefore, "sharing is an opportunity to release [people's cooperative] capacity, confined by competitive markets and bureaucratic states." So, if cities are shared creations with shared public services, streets, mass transit, and shared spaces, "truly smart cities must also be sharing cities" (ibidem). But, what does this mean nowadays when most people's livelihood, especially in the Global North, might depend on access to information, communication, and coding (Lash, 2002)? What is *commons* and what is the value of *commoning* in a city increasingly regulated by processes of data acquisition and exchange?

For Hardt and Negri (2009), the results of advanced capitalist production are expressed as a "collective intelligence" unlocked by the forced multiplicity and proximity that urban living implies. Value is created now by life itself, in mundane and very material practices of urban dwelling, social encounters, and social reproduction. Production extends to the city in its totality: it is the 'social factory 4.0.' As Negri and Hardt insist (2009, p. 154), the city is "a living dynamic of cultural practices, intellectual circuits, affective networks, and social institutions." In their view, the city is a social factory for capitalist accumulation and for the production of the commons. In other words, the city is "the *source of the common* and the receptacle into which it flows" since it produces, and is the result of, what they call "biopolitical power."

Many activists and scholars advocate for digital commons where ownership and control of data become a democratic practice of appropriation of technology. For instance, the "right to the digital city" (de Lange & de Waal, 2013) is centred on an alternative form of ownership of data, and the "informational right to the city" is grounded in a "deliberate project to re-appropriate and self-manage the information that we produce in a manner that we can both enjoy and can sustain" (Shaw & Graham, 2017a, p. 16). At least, citizens should have the right to understand what data are being generated about them, how these are compiled into information and the uses to which they are put (Kitchin, 2016b). Moreover, rewards from data ought to be socially shared or redistributed both within the public and within communities of practice (see Mazzucato, 2018a). Other solutions in this respect have included renewed calls for a form of basic social income which acknowledges the 'smart city' as a social factory, but also considers labour as a producer of use value beyond capitalist forms of production and appropriation (Monnier & Vercellone, 2017).

Partly acknowledging this, I want to highlight two major problems with digital commons: we can call them 'stewardship' and 'rent'.

A materialist critique of digital commons

Forms of democratization of data and software are not *per se* immune to falling into the "post-political trap of technological determinism" (McLaren & Agyeman, 2015, p. 201): there is the risk these initiatives foster 'commons' as a goal in itself, rather than 'commoning' as the process that leads to such a goal.

Therefore, some scholars prefer to put the emphasis on *commoning*, a set of practices which both "actively seek[s] to integrate resources from the state and capital into commons circuits" (Birkinbine, 2018, p. 291), and reproduces the commons through sharing resources in solidarity networks. Commoning shifts our focus on the long-term maintenance of the commons: "the true challenge of the commons," according to Huron (2015, p. 973). De Angelis (2017) argues that this hinges around the daily practices of the community of interest involved and around the bundle of rights attached and their enforcement, which should protect the commons and guarantee its reproduction. Thus, commoning is more than the social reproduction of life; it highlights the contested, open-ended, and political character of urban commons (De Angelis & Stavrides, 2010). At a deeper level, it is that moment in which social movements open up social structures to a higher level of political awareness, commoning is a "grass-roots project to build a new form of consensus around a different set of values and ethical codes" (Susser, 2017, p. 1). Internet equality claims do not stand as an exception since more connected people do not translate automatically into fairer communities, nor more liveable cities. This is because cities are messy and complex places and because infrastructures are sociotechnological assemblages that depend for their functioning on the practices, uses, and therefore skills of those involved around the milieu they foster. With algorithm-led technologies, this assumption is often subsumed to the process of acquiring data, selecting optimal profiles and responses, and enabling feedback – that is they seem to foster an elevated level of stewardship: doing things on other people's behalf, rather than commoning.

The second issue I want to highlight with regard to a materialist critique of digital commons concerns the actual process of social reproduction in cities. This critique centres on the 'right to the city' as the form of social reproduction which fosters the commons as the daily materiality of dwelling. With the financialisation of home and the transformation of use-values of dwelling into the exchange value of 'rent', the right to the city is subsumed to a very material category: 'rent' is, in fact, *both* the objectified value of urban commoning *and* the basis for its displacement. Thus, while Hardt and Negri (2009) call the new digital commons "immaterial" – such as knowledge, language, codes, information, competences, and affective practices (2009, p. 145) – each commons presents a new set of social relations and spatial organizations. I would contend that a materialist stance needs to be maintained with the notion that these forms of wealth are themselves produced and subject to capitalist forms of accumulation (see Harvey, Negri, & Hardt, 2009). As a consequence, they have not an anti-capitalist 'essence' deriving from the fact that they are produced *in common*, as Negri and Hardt appear to maintain (see Stavrides, 2016).

It follows that a materialist critique of digital commons (information, knowledge, data, software, etc.) ought to centre on daily practices of inhabitation and social reproduction, too. As Harvey et al. (2009) remind us, value is an immaterial social relationship which is, at the same time, objectified into the commodity and in the most material and ubiquitous form of 'commons', that is, money. In the process of urbanisation, 'rent' is one objective form of appropriation of today's commonly produced social value, the art of dwelling and living in the city. In fact, value increasingly manifests itself via the financialisation of 'home' as investment rather than universal human right; it is an engine for redistribution of wealth and an important *locus* for social reproduction (Aalbers & Gibb, 2014; Madden & Marcuse, 2016; Rutland, 2010). Thus, 'rent' has become an increasingly divisive form of commodity and accumulation, its costs occupying an ever larger quantity of families' income to the point that it affects directly life chances and lifestyles in the city.

While determining the appropriation of increased urban productivity to the benefit of few rentiers (direct displacement), 'rent' divides by promoting specific notions of urban commoning and by commodifying urban life further into lifestyles (cultural displacement). Moreover, 'rent' encloses urban commons, making it exclusive to some and excluding others (exclusionary displacement). I would agree with Bresnihan and Byrne (2015, p. 1) when they say, *contra* the immateriality of the commons: "for us the centrality of the commons stems from its immediate importance to social reproduction and therefore in our lives." Commons concerns, in fact, with a very material process of production and appropriation of socially produced value: bypassing a critique of the 'creative class' (Florida, 2003) is then unreasonable, since this narrates an intrinsic process of enclosure and exclusion at the heart of the 'smart city'. For me, the critique of 'rent' is central to the production of new urban and digital commons because it is central to the social reproduction of people living in cities, themselves workers in the 'social factory 4.0'.

It is through this agonistic, dynamic, and place-based lens that I understand public provision of the Internet. I advocate for a policy direction *alternative* to the prevalent neoliberal mode of making cities 'smart': this direction would pass through a process of municipalisation and democratic governance of critical infrastructures (such as data infrastructures and the Internet) and Public Commons Partnerships between the city, cooperatives, and social movements. The case study in the next section concerns the building and maintenance of a community-owned wireless network and connects the dots of a materialist critique of urban commons with the practice of commoning, attempting two conclusions. The first one is that 'smart' technologies risk having exclusionary effects on spaces of inhabitation, an urban *common*, through gentrifying pressure. The second point is that 'smart' technologies demand the deployment of a great amount of stewardship, crucial to the practice of *commoning*: there is nothing automatic about platforms and sensors in making communities 'operative'. Instead, these require intervention and help from community advocates with strong technical skills, ethical digital platforms, ethical Internet Service Providers, and ethical hackers, as I debate next.

Maintaining the Internet as a commons

In this section, I revisit an ethnographic account I wrote around a community wireless network in inner-city London called OWN (Cardullo, 2017), reflecting on its long-term sustainability through the notion of commoning.

Open Wireless Network started in 2008 from the rooftops of an iconic hackspace in Deptford, inner-city London.[1] It was a mesh of independent radios (nodes) which, by talking to each other and via ad-hoc gateways, provided extended broadband access to the immediate neighbourhood or passers-by in the reach of its wireless signal. Soon after its establishment, OWN peaked to over 100 nodes and 400 users at any one time (which at the time was considerable); however, in the last few years the project went through a period of decline since, with smartphone data network access as standard and a wider public provision in libraries and cafés, "some of the passion for independent Wi-Fi infrastructure building has fallen away" (Cardullo, 2017, p. 5). Informality and commoning were key to the initial popularity of OWN. This is because OWN resulted in a great value for its users, responding to the local population's real need: a working class and racially diverse neighbourhood where digital divides and the gentrification pressure of displacement are high. The mesh of nodes particularly suited transient and migrant people, students and temporary workers, but also less wealthy locals who do not have the capability to enrol in any official provision (for lack of residential documentation or income).

Community Wireless Mesh Networks (CWMNs) have been under development since the early 2000s thanks to the work of ethical hackers and activists responding to a series of local needs and generally advocating a more open, neutral, and democratic Internet (Cardullo & Roio (Jaromil), 2019; Medosch, 2003, 2015). Because of their inexpensive hardware and relative ease of assemblage, community networks have been deployed in many different ways, operating as specific solutions to local issues: therefore, each case is related to its wider digital and political ecosystem. First, they have been an answer to excessive connectivity costs, allowing communities to share the cost of network deployment and the use and provision of resources, such as community radios, applications and data, or listing of community events (for instance, both Ninux in Italy and Freifunk in Germany started in 2002 and grew to operate about 40k nodes). Second, adoption of CWMNs seems to have regained traction, especially in the United States because of the current Net Neutrality debate where Big Cable have been lobbying for selective throttling, slowing down, and metering, particularly in the Fibre-to-Home market. Third, community networks have been providing 'last mile' access to remote localities where the market has "failed" to deliver: this is how Guifi.net started in Catalonia in 2004, now bolstering over 33k operating nodes in much of Spain. Fourth, open-source hardware platforms like Arduino and Raspberry-Pi are deployed for low-cost and scalable mesh projects, generally for environmental monitoring in Citizen Science. Obviously, dense residential areas work better for crowd-sourcing since they can enable resource pooling and commoning across the network (e.g., file-sharing, game modding, environmental data). Fifth, the CWMNs has been occasionally tested by natural or man-made disasters, for example, when Hurricane Sandy destroyed most of the communication infrastructure in NYC, Red Hook mesh functioned as an effective back-up in the area.[2] Finally, community networks are deployed to fight data extractive practices and surveillance, since mesh are, ultimately, Intranet systems with locally controlled circulation and repositories of data. This is the case of SNet in Havana, a wide and expanding mesh where about 20k users play networked games and exchange electronic items.[3]

As with many other public hotspots, there was a problem affecting the provision of OWN. One of the hosts of the mesh recalls that, "because of the speed involved, OWN is not good for videos, but it is for general browsing" (cited in Cardullo, 2017, p. 4). The ability to allocate resources, bandwidth, and speed, remains important as the network grows. For instance, OWN got its broadband provision from a higher hierarchy level, Tier 1.[4] A more robust bandwidth allows in fact a higher number of users connected at any time. It also enhances strategies of connectivity across neighbourhoods where high-rise buildings risk blocking the wireless signal. This reminds me of the Chattanooga case: speed, bandwidth, and cables matter sometimes, as it matters that the provision of the Internet is via a trustworthy public infrastructure, the 'backbone'. This could be provided to ethical ISP and cooperative networks at subsidised rates, as the case study of Queensland, NYC, revealed.

While the detailed ethnography of OWN and its community members can be read from my previous article (Cardullo, 2017), the research on the development of Open Wireless Network (2013–2015) reveals two issues important to this chapter: 'stewardship' and 'rent'.

'Stewardship'

Maintaining a mesh of hardware, patching the software, and training people to use it was possible only with a large amount of stewardship from trusted community advocates who themselves had strong technical skills. We can call them 'ethical hackers', people able to mobilise a good degree of social and techno capital, time for mostly voluntary work, and some occasional funding. Indeed, the major challenge for community networks is their long-term operation and maintenance, especially as many more lay practitioners get involved when the mesh tends to scale horizontally and vertically (as geographical extension and number of users). Technologies we take for granted in our everyday practices demand in fact induction, participation, and care.

This is where training and support became strategic during the OWN project, enabling a bond dictated by practice. The knowledge transfer generated during training sessions and the social capital that training produced were crucial to the project: "OWN was to expose the idea of mesh network in a way that people would get *an experience* that was both practical and informative" (James, founder of OWN, in Cardullo, 2017, p. 6). "We didn't know about the 'smart city' and stuff like that," says another node owner, "we just did it because it felt right" (ibidem). This speaks to me of a gift economy, an attitude towards helping others, and communal maintenance *before* and *beyond* the 'smart city' discourse around Living Labs and digital innovation districts.

Community Wireless Mesh Networks have been experimenting with different models for sustainability and governance, the majority relying on voluntary work from activists and contributions from their members to offer Internet connectivity (Cardullo & Roio (Jaromil), 2019). Usually, node owners buy their hardware: radio, antenna, cables (e.g., Freifunk in Germany, Ninux.org in Italy, S-Net in Havana, OWN in London). Guifi.net in Spain, instead, has been asking members (including commercial operators) to adopt a specific license, subscribe to an arbitrator agreement, and contribute with a fee which includes upgrade and development of the network as well as its maintenance. This is carried by senior members, professionals who also offer other telecommunication services over the network (landline telephone or IT support), guided by "clear ideas and strong leadership" (Baig, Roca, Navarro, & Freitag, 2015). Finally, Red Hook mesh in NYC offers a model where installation and maintenance are provided by Digital Stewards: these are young residents from the local public housing employed in a paid fellowship (20 hours a week at $8.75 an hour), funded by a public grant (Cohen, 2017).[5]

Whatever the model adopted, every group involved in Community Wireless Mesh Networks maintains a local weekly or monthly assembly for face-to-face discussion and problem-solving: for OWN, in London, this was a free training space called 'Wireless Wednesday'. This is because making community operative is an endeavour rooted in social trust, which is a long-term relationship involving at least two things: a place easy to recognize, in the locale where people actually live, and projects that are engaging because they are deemed *useful*, that is, they are perceived as *doing something* for their users. Provision of the Internet for people and 'things' requires medium level stewardship and skills; however, stewardship becomes essential with regard to Community Wireless Mesh Networks because the higher the involvement of lay people, the harder it will become to include them in the development and adoption of the technology. In fact, smart technologies demand the deployment of cultural and social capitals because they are linked to social exchange and their implementation is conditional to contextual arrangements in communities of interest and localities: that is, there is nothing automatic and deterministic about platforms and sensors in making communities "operative."

'Rent'

The second important issue that my research on Community Wireless Mesh Networks highlights is that these initiatives need to be evaluated in relation to the dynamics of urban space and 'rent'. To my mind, the problem that wireless networks like OWN face is the coming together of two opposite forces. One direction pulls towards the 'local' – sharing resources and data, and bandwidth and bulletins, usually within the limited reach of the wireless wave. But the 'local' is not just the *locus* for direct involvement, neither is it here intended as opposite to the 'global'. Importantly for the functioning of OWN, the local is where a gift economy of exchange and its expectations materialise.[6] This gift economy is expressed via the sociality of the mesh, for instance in relation to outreach of new hosts and negotiation of bandwidth provision. In this sense, for me, the 'intelligent city' passes through the dis/organisation of infrastructures of communication, which takes into account equality and freedom of access, privacy and security rights, knowledge transfer, and ability to choose software solutions.

The other force is, instead, centrifugal. It seems to tear apart that neighbourliness on which the Community Wireless Mesh Networks rely. This is because of the ongoing privatisation of residential solutions for new upmarket buyers and the consequent displacement of working-class residents. OWN started developing alongside a changing urban space: a gentrifying neighbourhood implies geographical displacement of people committed to the cause of the commons – hacktivists[7] but also, in my focus, its working-class residents who were the main beneficiaries and users of the mesh. Additionally, gentrification brings forms of cultural displacement (Marcuse, 1985; T. Slater, 2009) since gentrifiers boast new attitudes, models of consumption, and lifestyle expectations which sit at odds with the politics and practice that networks like OWN delivered.

In fact, the paradox of 'proximity' – wireless networks are necessarily territorial – in a gentrifying neighbourhood puts at the centre of the organisation and main- tenance of Community Wireless Mesh Networks a slightly different notion of the 'intelligent city'. From my field observation and findings, I derived a strong sense that new luxury flats can limit the outreach efforts for new nodes. This is for two sets of reasons. First, because the physical city now boasts more exclu- sive enclaves. Second, and more importantly, the gift economy on which OWN was based might be negatively affected by individualised lifestyles and aesthetic consumerism, proper of the neoliberal subjectivity. OWN is in fact based on a gift economy made of shared broadband as well as ongoing maintenance of soft- ware and hardware. To my mind, this circulation is made of actions as well as affect. In their study of working-class personhood, Skeggs and Loveday (2012, pp. 475–476) invite us to think of 'value' not just in economic terms (accrual of various forms of capital), but also "relationally, as a more general ethos for living, for sociality, and connecting to others, through dispositions, practices and orienta- tion." I would argue that OWN contributed to building and circulating a certain type of *experience* that is rooted into the social fabric of working-class Deptford: "OWN is about local people who give a bit back to other local people in the area. I have been here for 12 years . . . there is not much money in the area, you know," says another host of the node (in Cardullo, 2017, p. 11).

These two issues at the root of a community-owned Internet network, 'steward- ship' and 'rent', suggest to remain vigilant towards enthusiastic developments of bottom-up and citizen-centric 'smart city' initiatives. Conversely, it is tempting to welcome the coming together of public investments in critical infrastructures and the communitarian ethos of sharing and caring. Although the mechanism of governance needs to be evaluated case by case (see De Angelis, 2017; McLaren & Agyeman, 2015; Morozov & Bria, 2018), this novel configuration would re-propose the role of the public as a competitive stakeholder beyond the neoliberal rhetoric of the 'market failure' (see Mazzucato's argument for the "public value," 2018). At the same time, it would preserve ethos, autonomy, and some degrees of infor- mality proper of grass-root organizations.

A mixed approach in the provision of the Internet – which I use as an entry point on the critique of the 'smart city' but it is not, by any means, conclusive or exhaustive of the debate around it – may be able to satisfy few of the categories above, by way of including: grass-root initiatives (in the forms of co-operatives, citizen groups, and social enterprises such as small and ethical ISPs); sharing of networks and licence spectrum (the latter being a commons too, rather than a commodity);[8] public engagement from the city (as an organization with social and political goals); and an adequate level of public investments (not only in place of 'market failure'). More generally, I think of democratic and inclusive governance as a double-faced Janus which offers benefits to both city and communities. On the one side, it presents itself as an essential element for commoning practices, which would involve citizens as equal stakeholders with rights and entitlements rather than as consumers of a private utility provision. But on the other side, democratic and inclusive governance becomes also necessary to the city that takes part in the *effective* demands of its citizens: no Amazon delivery robots or drones here, but rather more affordable and reliable bandwidth! As Stavrides (2016) writes, any commons is held in a dialectic relationship with its users because "common space keeps on producing those who produce it."

An intelligent city?

We can reframe the debate on the commons from the classic triad Private/Public/Commons to the emerging configuration of Capitalism/City/Commons (see De Angelis, 2017). Here, the struggle for appropriation of value moves iteratively through the two circuits for creation and circulation of wealth and well-being: the circuit of capital and that of the commons. At the same time, the city becomes a flexible scale of reference – e.g., from the federal scale of the 'backbone' and spectrum governance to the neighbourhood scale for the 'last mile' provision.

It is at the city scale that the *Commune* comes to life. Harvey maintains that "through their daily activities and struggles, individuals and social groups create the social world of the city and, in doing so, create *something common* as a framework within which we all can dwell" (2011, p. 104). Cities have a lot to learn and gain from nurturing urban commons – the spaces of solidarity, social reproduction, and mutual aid – through ethical procurement and cooperatives and collectives of people pursuing goals other than the private profit in public service provision. The city is, in fact, the most complex form of human organisation which manifests itself as a commons by the way of "experiencing collective work, among strangers, to govern non-commodified resources in spaces *saturated with people*, conflicting uses, and capitalist investment" (Huron, 2015, p. 977, emphasis added). Common-focused institutions would maintain, for example, "trust on trustless networks" (Schneider, 2018), by fostering transparent governance. This has been exceptionally lacking in networked and digital societies after decades of neoliberal governance at distance, marketisation of public services, individualisation of social responsibility, exasperate nudging behaviour, and overarching algorithmic control and surveillance.

Conversely, as Henry Lefebvre (1996 [1967]) suggests, urban space is the "work of art" of its users: it is appropriated by the everyday practices of the people who inhabit it. In this guise, the right to the city is "a right to change ourselves by changing the city more after our heart's desire" (Harvey, 2003, p. 939). This is a space ready to accommodate citizens' political claims: according to Marcuse's (2009) reading of Lefebvre, this is the most radical idea he brings forth because it opens both to those people deprived of material subsistence and legal rights (*a cry*) and to alternative futures (*and a demand*). Thus, *commoning* happens in the localities of the everyday, the space of inhabitation, in and through urban space. In other words, pursuing the 'right to the smart city' means creating grassroot cities that are inhabited and lived, rather than rooted in and driven by technological capitalism and the solutions this fosters (Kitchin, Cardullo, et al., 2019).

And yet, according to some commentators, the great achievement of algorithmic-led apps that allocate goods and services in the platform economy is the emergence of a new form of trust horizontally established between the two sides of each market: for instance, between the drivers and the passengers engaged in an exchange through the Uber platform. Filtered by algorithmic rationality and calculations on the basis of utility, time, ratings, and previous feedback, consumers and providers maximise their utility because – so the neoliberal rhetoric goes – they have all the relevant information to make an informed and rational choice. Thus, the digital platform becomes the epitome of a perfect market which allocates resources via optimal prices: you cannot get a better deal, whether you sit in a bar waiting for your lift or in a car waiting for your passenger. This new form of trust would replace, it is argued, the vertical trust provided by institutions, such as governing bodies, licence agreements, work contracts, and any middle-man activity.

This is, however, a hybrid form of trust, partly between agents responding to other agents' feedback and choices and partly between algorithms on each other's devices which respond to *the same* platform's parameters and programmed scenarios. The practices and uses of those involved are indeed subsumed to the process of acquiring data for private and corporate use; or producing optimal profiles and responses which enable selective feedback. People and things are actuators of the data exchange through networks, continuously performing acts of trust at the margin of the mesh made of interconnected devices and sensors. But while algorithmic trust happens between long alphanumeric keys which recognise each other and perform a so-called 'handshake' in milliseconds, social trust hinges on long-term relationships and slow exchanges performed mostly face-to-face and in the spaces of the everyday. Social trust builds over time, upon frequency, social bonds, and culture, and around social spaces shared in commons. This is where ethical hackers and community organisers with strong technical skills can help mobilise knowledge transfer and limit digital divides, and for people to get actively involved in the making or maintenance of technologies in social space. Most of the technologies currently deployed in the 'smart city' are, however, beyond people's agentive interaction with their computational processes enabling black-boxed, autonomous, and automated responses. The gap between algorithmic trust and social trust is partly responsible for the fact that platform economies and 'smart' technologies have exclusionary rather than inclusive effects on people's participation in their design, decision-making process, and governance.

In addition, by making decisional processes and governance more dependent on algorithmic choices, there is the risk to force out of informality established networks of subsistence. This is particularly relevant to researchers investigating informality in the Global South where the informal use of scarce resources is thought to display a great deal of collective intelligence, making it possible for people to reproduce their lives beyond the coercive, regulatory, and controlling gaze of existing institutions and the market. Although meant usually as an approach to urban poverty, informality proliferates in the cities of the Global South at all levels, among dwellers of informal settlements and practices of social reproduction, economic transactions, and daily getting by (Roy, 2011). There are concerns that the 'smart city' will work towards formalisation of informality and, thus, have pernicious effects on those people who make their right to inhabitation based on informal relations and on sharing livelihoods *in commons* (Smart, 2018). For instance, 'smart city' initiatives that aim to eradicate informality have been rolled out in India, such as de-monetisation of currency, systematic harassment of street vendors, and the introduction of Aadhaar (an identity and service card based on biometric and demographic data) (Datta, 2018).

How can an 'intelligent city' position itself with regards to these many external-ities and exclusionary effects for its citizens? We have seen many responses that cities around the world have started to implement in order to protect themselves, their budgets, and their citizens from the wave of enthusiastic technologists, indus-try representatives and, sometimes, academics who propose to 'fix' entrenched urban issues. Some cities have started to take the matter of Internet provision into their own hands and away from the market. Some cities have adopted tougher regulations for the platform economy. Other cities have encouraged the prolif-eration of cooperatives. For instance, Barcelona City Council started a program dedicated to developing platform cooperatives in 2015 which provides matching funding to support cooperative entrepreneurship, as well as half of the funding for privately controlled infrastructures to be used in the development of publicly or cooperatively controlled businesses. The cooperative movement is a sector tradi-tionally devoted to sharing and to the provision of services to the person, to creat-ing systems of mutual support between members, and to maintaining solidarity networks for the have-nots. Cooperatives have helped millions buy homes, and enabled favourable access to credit for the many. They have been traditionally community-oriented, for instance, pioneering the organic revival and the means of delivering locally produced food. One way to answer to the exclusionary risks brought in by privately owned platform economies and poor governance is, thus, to look at how the cooperative economy is responding to the digital transforma-tion and their relations to the city (Scholz, 2016).

More established cooperatives started to adopt 'smart' technologies to optimise their service and to meet an audience more keen to use mobile apps. It is the case, for instance, of CO.TA.BO (Cooperativa Tassisti Bolognesi) which started in 1967 in the Emilia-Romagna capital, in order "to provide a better public transport service for the city".[9] The coop has eventually developed its own app, TaxiClick Easy, which has the same functionalities we would expect from the Uber app. One main difference from the Silicon Valley 'transport company' is that the coop maintains its traditional ethos of mutual support and fair retribution for its associ-ates and high ethical standards for its users. Traditionally, the coop has guaranteed to its drivers admin and legal service, insurance and maintenance for the vehicle, and even the petrol, which is provided by their own infrastructure and costs less than on the open market. Moreover, the cooperative and its data are owned by the workers and/or service users.

As seen, McLaren and Agyeman suggest (2015, p. 1) that "good governance and collective city structures" are what successful cities need. Barcelona and

Bologna are good examples of this, and it is not by chance. Both cities have a long tradition of community engagement, strong civil society, and established cooperative economy. Barcelona has a long history of urban movements and strong neighbourhood self-organisation, dating back to the First Spanish Republic (Nel·lo, 2015). Bologna cooperative economy is well-developed and, more importantly, embedded in the cultural fabric of the city (and of the Emilia-Romagna region). Another example is the recent pilot Fairbnb, an ethical coop that links many cities including Bologna and Barcelona. It promises to foster transparency and legality in the sharing rental market, with their 'one host, one home' policy and payment of taxes to the local authority; moreover, 50% of the commission is going to fund local community projects; finally, being a cooperative, the organisation ranks are open to the community of hosts and services who want to become members of the coop.

On the Internet provision front, US cooperatives started providing answers to the quasi-monopolies and poor service offered by Big Cable: in January 2019, there were over 800 initiatives, of which 500 are served by some form of municipal network and more than 300 are served by a cooperative.[10] "A cooperative internet might seem utopian," says Nathan Schneider (2018), "but cooperatives brought electricity to rural America when no one else would . . . Next, the Internet. We have done this already, and we can do it again."[11]

Further, radical experiments in techno-politics start shaping the debate on citizen participation in the 'smart city', with some cities adopting open source digital platforms as a form of democratic governance and citizens' participation in the decision-making process. Techno-politics is much more than e-government, such as online city services and feedback. It aims at making tactical, strategic, and critical use of digital technologies for collective political action, with a focus on improving democratic practices in order to advance participation and decentralization. This is the case, for instance, of the successful *We decide* platforms, designed and first adopted in Madrid (*Decide Madrid* in 2015) and Barcelona (*Decidim* in 2016) and now exported to many other cities: typically, this sort of platform boasts discussion threads, scoring, ranking, file sharing, event coordination, thematic clustering and visualisation, notifications about issues or themes, and digital space where citizens deliberate and directly decide on proposals, budgets, and plans for their city. As for the city of Barcelona, the platform counts more than 14,000 proposals and over 1,000 citizen initiatives, enabling 32,000 people to participate in the formulation of strategic city plans.[12]

Partly acknowledging this, I would warn about the risk of focusing on the outcome (open data, free software, number of clicks) rather than the process which maintains and reproduces such commons. Techno-politics advocates (e.g. Smith and Pietro Martín, 2020) are clear that their approach works best when social movements and political institutions commit to developing the technology in democratic form, such as the 15-M anti-austerity movement and the following political platforms (Barcelona in Comu', Podemos and Ahora Madrid) proved crucial for techno-politics ideas. The participatory phase, in fact, is meant to involve always an active consultation process: for instance, over 700 public meetings were organised around the strategic planning of Barcelona city (Barandiaran et al., 2017). Moreover, the pace of the administration machine and the scale at which decisions on public land issues effectively happen hinder the automatism of technology-enabled decision-making processes.

From the above discussion, it derives that forms of governance enabled and aided through digital platforms require modalities of operation that are not autonomous from traditional decision-making processes or lobbying and trust building political practices; neither are they automatic. That 'not everyone knows everything' is almost a truism. But there would be a generally common sense around fixing a bike wheel or turning the water tap off. These long-established technologies ultimately present visible screws, bolts, or switches. Most 'smart' technologies, instead, are sealed in their soldered plastic containers, or they present firmware blocked by encrypted access codes and software closed by proprietary keys. They are black-boxed or accessible to only a few technology skilled people, amateurs, or hackers. In commoning these, intervention and help from community-focused advocates with strong technical skills is probably always necessary in order to start thinking in terms of an 'intelligent city', a city that is technological *enough* but with a strong focus on citizens' rights and needs. These advocates are going to provide the missing links in the commoning infrastructure; they are likely to be the people behind ethical digital platforms such as FairBnB, or behind ethical ISPs, with strong principles around digital rights and affordable Internet provision; or they are 'ethical hackers' with strong local bonds and an eye to social justice (something similar to the mechanics, the admin/legal team, and the affordable petrol available to the drivers of the taxi coop in Bologna).

In other words, it appears to me that 'smart' technologies come to the price of an increasing stewardship. This is certainly due to maintenance issues around the infrastructure or the 'smart' device, as these need skills and digital literacy to operate. At best, stewardship means an attempt to reduce the 'digital divide' due to people lacking in resources the 'smart city' demands of 'smart citizens': techno, political, and social capitals. Within a sociotechnological landscape of increased stewardship then, it appears crucial to me that stewardship circulates within a well-routed cooperation economy, driven by values of solidarity and mutualism and by the respect of rights and the common good.

Concluding remarks

From this chapter discussion, I take that an attention to *commoning* would highlight the exclusionary practices and the forms of capital demanded from ordinary people in order to meaningfully participate in the 'smart city' at different points in the data infrastructure 'stack'. For instance, the political battle for a universal decent broadband service risks being flawed by perpetuating practices of data extractions for the private goal. Equally, advocates of data commons might be at odds when explaining the use-value of urban dashboard in relation to real-time city governance and surveillance. The gap between algorithmic and social trust is a critical issue through which I built a materialist critique of 'smart commons' (information, knowledge, data, software): in this chapter I started asking, what does the 'right to the smart city' actually mean vis-a-vis systems and structures of inequality (re)produced in the 'smart city'?

Cities have a lot to learn and gain from the commons. An alliance between city and the commons, rather than the market, can start removing some of the barriers in the adoption of technologies which depend heavily on the Internet of People and Things. Rather than an heroic immersion in the back alleys of cybernetics – focusing on protocol exchange, encryption, and blockchain – a pragmatic approach to commons in the 'smart city' is required. This approach would consider, for instance, that the myriad of citizen-science and crowdsourced projects and communitarian networks suffer from conspicuous stewardship and long-term maintenance, resources, and scalability issues, which maybe cities can support by taking the Internet backbone under their own capacity or favouring municipalised or cooperative enterprises, such as small and ethical ISPs. Conversely, there is an uncertain role of the city in drawing policies that support (or hinder) the inclusion of communities, devolve (or take) power to (from) citizens, and enact (or re-centralize) forms of open and democratic governance.

Notes

1 Recently closed after H&S and Fire regulation checks by Greenwich Council, http://wrd.spc.org/described/deckspace/
2 However, since CWMNs operate in unlicensed frequency bands which are subject to uncontrolled interference from a range of sources and overcrowding, their efficacy during emergency is disputed.
3 Snet is not offering a gateway to the public Internet since connectivity in Cuba is patchy and limited to public parks and schools because of the US embargo and restrictive local policies.
4 https://redrawinternet.com/internet/
5 www.nytimes.com/2014/08/24/nyregion/red-hooks-cutting-edge-wireless-network.html
6 This tension of proximities and divergences – an open, but off-line network – is rendered in OWN users' involvement with their local social landscape.
7 See Medosch, cited.
8 https://www.apc.org/en/news/whats-new-spectrum-lets-make-sure-we-can-use-it-what-needed-conversation-peter-bloom
9 www.cotabo.it/?lang=en
10 Institute for Local Self-Reliance, https://muninetworks.org/communitymap
11 https://ioo.coop/2018/02/next-the-internet-building-a-cooperative-digital-space/
12 www.decidim.barcelona/

9 Conclusion

Do we need the 'smart city' after all?

While editing this manuscript in confinement from my flat in Italy, the coronavirus (COVID-19) crisis exploded in all its dramatic social, economic, and political violence. What has been absolutely clear is that this crisis is not just health-related but it has much deeper and long-lasting effects on every country, sector, and social group. Stepping back for a moment from the tragedy of many, especially elderly people, genuine worries of a different nature come to my mind: as entire cities and nations have decided to lock down their citizens and hospital facilities turned out to be insufficient, the pandemic appears more and more as a crisis of urbanisation, mobility, globalisation, federal projects, and neoliberal economics. Moreover, the crisis intertwines deeply with 'smart' technologies – an 'urban' crisis is also a 'smart city' crisis – giving me a chance to sum up a few themes I dealt with in the book. I want to put forth four sets of considerations around what has been a sudden, evolving, and hard to pin down global emergency. Although rooted in the reasoning developed so far, the considerations are conceptual at this stage, rather than empirical, due to the contemporaneity of the events and the impossibility to run out fieldwork in the present conditions. They open to research questions and issues around the 'smart city' and digital society of the future.

Super surveillance and the experimental state

At this moment, the first and more pressing consideration comes from a big worry. Keeping an entire population at home, mapping the epidemic's movement, and grounding most of capitalism production are not policies to be taken lightly. These exceptional measures, justified by a state of emergency, however, demanded the surprisingly minimal use of force. Rather, these policies demanded the management of the disciplinary, controlling, and biopolitical gazes of the state, an almost panoptical and totalitarian presence of authority supported by the broader deployment of technologies for the regulation of people's daily life, by the dashboarding of almost everything, and by nudging into individual compliance.

Thus, with the urge of experimenting a new pharmacon able to contain the virus, states have also been experimenting with new social strategies of discipline and control, coupled with neoliberal techniques of monitoring, tracking, and dashboarding of individual health and global movements. Various technologies are being used, such as:

- drones to check on people's gatherings (as in Naples, the area of highest people density in Europe);
- monitoring apps, supported by data contagion reports, to control the movement of the epidemic surge and people directly affected (as for COVID-19 Tracker app for I-phone which is a "solid disease tracker app that displays the data from the World Health Organisation" or a Close Contact Detector app which "will warn you if the person sitting two rows near has been diagnosed with Coronavirus or not");
- modelling of epidemic curves with stochastic predictions (as in the UK with the so-called 'herd immunity');
- thermometer-enabled helmets and thermo-scans for border and patrol police (almost everywhere while the tourist and flight industry came to a halt);
- publicly consensual confinement of people tested positive to COVID-19 through mobile phone tracking (as in South Korea which deployed live GPS tracking of the patients who are at home in quarantine and those who have tested positive, using also data from surveillance camera footage and credit card transactions) (see Kitchin, 2020 for a detailed overview and critique of these apps).

These exceptional measures confer the normal deployment of the state's monopoly of force to otherwise experimental devices and techniques. In the face of the state of emergency, there appears to be a more pronounced suspension of privacy and other civil liberties claims with patients', or symptomatic people's, medical history becoming of public concern.[1] Further research will need to provide more accurate and detailed assessments of how medical data have been, and are going to be, used from now on. However, the totalitarian affordance of these new technologies has become painfully evident: a mobile phone is, *de facto*, an electronic bracelet placed on people's ankle! It remains to be seen when, and in what measures, such exceptional and experimental techniques of governance will be reverted or eliminated. For the moment, my understanding is that 'smart' technologies being deployed in cities as a consequence of the pandemic – dashboarding techniques, online mapping, DNA mapping, tracking of populations' movements, smartphone-enabled apps, devices for the patrolling of borders, helmets equipped with thermo-scanners, digital thermometers, and sophisticated face-recognition cameras – are all playing an increasingly invasive role in people's lives.

Never ending technological solutionism

At the highest moment of this global emergency (the United States has registered the largest numbers of contagions in the world and the world's biggest democracy, India, is officially in 'lockdown'), Huawei's president for Western Europe tweets: "5G will enhance the effectiveness of pandemic prevention and treatment, and drive the digital transformation of healthcare systems in response to major public emergencies." Unfortunately, the much- needed white paper accompanying the tweet was soon taken down, maybe because of the broad protest the tweet sparked. The public display of arrogance from the High Tech industry has not gone unnoticed by the public. As I have unpacked throughout the book, the 'smart city' discourse is winning according to the belief that technological solutions are applicable everywhere (technological determinism), thus becoming an unquestioned policy goal. This discourse, it is worth repeating, concerns the adoption of technological devices, apps and sensors, and the massive deployment of data deriving from these, in order to 'fix' urban issues and safeguard all aspects of the life of citizens (technological solutionism). The recent rise in the availability of big data in every sector of city governance and the lucrative tags attached to the management and control of such data have increased the demand for data analysts, software engineers, and creators of urban models. A crucial step in the making of this 'smartmentality' (Vanolo, 2014) has been the coming together of different forms of expertise, often the outcome of academic and industry joint ventures supported by supranational public funding strategy. These professionals too are massively feeding the 'smartmentality' trope: that is, the belief that every aspect of city living is measurable and, thus, predictable; and that an algorithm-based app or device is more able to understand the pulse of a city and to foster a solution.

In other words, the 'smart city' remains first and foremost an ideological construction, a cultural trope of speed, growth, and modernity. The scope of critical scholarship, then, is to 'provincialise' such a trope and make it strange. Different questions might need to be put on the table, for instance: Why would cities retrofit pavements and urban furniture in order to accommodate Amazon delivery robots and driverless vehicles? What are the social costs of massive technological unemployment in the delivery and driving sectors? In the face of the looming pandemic and the stark social and economic crises this has exacerbated, futuristic and technocratic discourses seem to me incredibly lame. The 'smart city' as a unified ideal starts appearing to some as a less solid proposition than initially thought, with city leaders and civic society groups voicing their concerns, doubts, and policy alternatives to city managers, academics, and industry representatives. In other words, the bullshit seems to have come to the fore.

The IoP strikes back

As a consequence of the fast spreading of the contagion, the Italian government, followed now by many other countries around the world, urged to lockdown the entire population: "Io resto a casa" (I stay at home) is the new buzzword bouncing back from ministerial ordinances, social media hashtags, and flashmobs of celebrities to cars or drones equipped with loud speakers. Schools and universities have been closed, too. As a consequence, social media, teleworking, online courses, chats with tutors and Zoom (or similar) video calls with entire classrooms are becoming, slowly and patchily, the new norm of these troubled days.

To this regard, the Italian Department for Innovation and Digitalisation has launched an initiative which aims to reduce the negative impacts of working from home and learning at a distance. Under the banner of "solidarietà digitale" (digital solidarity), hundreds of big and medium tech players are offering services and discounts on their digital products, such as: extra gigabytes for home entertainment, free online access to magazines and ebooks, discounted server storage, free certified email address ("to avoid going to the post office"), and of course a wealth of distance learning facilitations such as online training, digital platforms and solutions for teleworking and distance conferencing. Similarly, the US Chairman of the FFC, the federal agency responsible for implementing and enforcing US communications law and regulations, pledges to keep Americans connected by calling on private Big Cable "to relax their data cap policies," on telephone carriers "to waive long-distance and overage fees," on those that serve schools and libraries "to work with them on remote learning opportunities." Indeed, in the digital solidarity race, the market does not seem to react automatically and accordingly. Free Press, for instance, calls for Internet Service Providers to play their part for the public benefit: "Especially during a crisis, Internet and phone access should be accessible public services like water and electricity."[2]

One of the pillars of the post-Covid society, as this is imagined through political discourses and unusual public spending, is then to build a strong 'digital society'. But, on whose terms? While these solidarity measures are certainly welcome in a dystopian climate (the emergency is seriously affecting every worker and learner in the country!), I believe there is a need to evaluate critically, and in the light of our discussion on the 'smart city', what this means in terms of the digitalisation process of the everyday, the Internet and data infrastructure, and the 'right to the smart city'.

First, the 'digital solidarity' campaign has made clear that digital infrastructures are a pressing need in many fields: online working, leisure, and communication activities are an increasing demand on current lifestyles. Although exceptional, there will be a booming effect on the digital sector and online business in the long run, forcing digital practices and learning onto people and institutions that have resisted this change thus far. As a consequence, I think there is more than ever the need to boost the Internet of People (IoP) for the many.

Second, the 'digital divide' has turned this technological 'experiment' into a social issue: in the United States, for instance, approximately 22% of households don't have Internet at home, including more than 4 million households with school-age children, and 8% of households who have the Internet rely exclusively on mobile broadband,[3] and this has become increasingly a racial digital divide (Turner, 2016).

Third, the 'right to the smart city' needs also to take into account the data and Internet infrastructure as strategic. Universal access to this infrastructure ought to be priorities if we were to consider equality and fairness as driving principles of digital societies. It follows that such an infrastructure cannot be left to private and unregulated oligopolies, as it is at the present almost everywhere, but rather it should be organised and delivered by municipalities as a public service for the common good. This alone is not going to challenge the many issues of digital capitalism: a whole set of services, platforms, data centres, clouds, and infrastructure governance would need to become regulated and/or brought into public hands, whether by the state (for larger infrastructure build, such as the 'backbone') or municipalities (for the 'last mile' delivery and local governance). However, a municipal and public Internet and data infrastructure may provide a gateway to increased local control and accountability.

Fourth, it seems to me that digital capitalism is responding to the crisis with a vested interest which leaves unresolved the pressing needs of a population forced to stay 'at home'. That giant providers such as AT&T remove caps from their users only made it clear that such restrictions are not dictated by technical needs of efficiency, but rather are unfair accruing of profits by private digital companies. That phone providers like Iliad offers 10 gigabytes more per month to their users in Italy is just a drop in the ocean: to put this in context, a film in 4k which is the high-quality standard of today's streaming uses almost as much as bandwidth as the suggested 'gift'. It seems to me that more ought to be done, especially considering the long-term effects that such generosity will eventually generate: more accounts and more private metadata, more users' details and eventual increase in subscriptions.

Moreover, the same institutional sites promoting the digital solidarity campaign have been quick to warn against scammers or hoaxes, and poor services hiding marketing messages. For instance, Google has been under fire by US Senators for continually providing ads on masks and hand sanitisers to users searching "coronavirus" and similar items, also engaging in "price gauging".[4] The window-dressing effects of such measures are likely to generate more profits in the long run for these digital giants to the point that such a 'solidarity' campaign looks to me more akin to a marketing opportunity hidden behind a humanitarian face.

Communitarian response and the 'public'

The prolonged lockdown and the stop to most industry and services deemed 'not essential' have generated predictable economic emergency for many people. Solidarity and ethics of care, not unlike the virus, feed on human contacts, face-to-face understanding, and long-term trust building: practising solidarity when you have to practice 'social distance' can therefore sound like an oxymoron. What will the cities of the future and commoning practices look like after the pandemic? Are solidarity and commons going to appear prevalently in a digital form?

Evidence thus far suggests that forms of power – in the sense of social infra-structuring, solidarity networks, commoning practices and new forms of protest, contestation and socialisation of political struggles, but also cash transfers with redistributive effects in the forms of Universal Basic Income, and finally re-jiggling of public space in favour of leisure time and pedestrians – have migrated not just from private money to the state, but from both market and state to the commons. Amid difficulties and lockdown directives, we are witnessing growing protests and political activism through social movements, but also mutual aid commoning and solidarity response from civil society, associations, volunteering sectors, and normal people.

The pandemic seems to have accelerated the process of social recomposition between social movements and the commons, thus responding to the aggravated conditions of social reproduction and inequalities which have been perpetrating at least since the 2008 financial crisis (see Massimiliano De Angelis & Shannon Mattern's recent interventions about this).[5] Indeed, in the last days we have witnessed the growing demands from sectors of civil society deemed now 'essential' and the lining of the same (mostly black and brown) bodies on the line of protest for their human dignity recognition over racialised policing. Conversely, solidarity responses from civil society, associations, volunteering sectors, and normal people have increased the number of food banks, organised shopping service for elderly and people in isolation and distribution of first emergency items including baby food and toiletries, or drop off of medics from hospitals to their homes for free, and a myriad of other neighbourhood lines of help. While Big Tech has been showing to pursue opportunistic ventures, the not-for-profit approach of Living Labs and ethical hackers, in Italy and elsewhere, have been producing much-needed oxygen masks by way of converting a commonly used diving mask (thanks to their own 3-D printed kernel).[6]

Amid understandable difficulties, in my hometown volunteers have been visiting people in need and delivering goods to their homes. The service was first initiated by a community group and then taken on board and coordinated by the Social Services. The city listened and responded to people's needs, coordinating with volunteers, supermarkets, and private donors the commoning side of this social infrastructure. It seems that servicing the most in need and supporting communitarian chains of solidarity have become a 'public' issue again, deserving a public response. Indeed, the Italian government made a substantial (although insufficient) transfer of cash to cities, recognising the crucial role of municipalities in handling this unprecedented crisis. Barcelona immediately suspended rental payment on city- owned proprieties and urged private landlords to do the same.

The notable elements here are the inversion of the trends from years of neoliberal policy of austerity and cutting in public services, and the fact that municipalities are recognised to represent the shared common ground on which everyday life happens. The current crisis offers perhaps a good chance for ditching neoliberal urbanism, for ending the competitive push of cities against each other, rather for entailing cooperation and solidarity between municipalities, and indeed between the city and its citizens. An 'intelligent city' can only work with regard to the increasing (digital and material) needs of people and things in the context of globalised economies and localised solutions.

Towards an 'intelligent city'

The future 'smart city' is up for grabs, limited – according to the business guru, Klaus Schwab – "only by our imagination." On the same tone, the recent 'smart city' platform organised by the European Commission in Brussels has been inspiringly called *DREAM (Demonstration, Reinvention, Engagement, Adhesion, Mobilization)* because, "All innovation begins with a shared dream."[7] In 2019 I was at the smart city expo in Barcelona (SCEWC) and the marketing slogan was "Cities made of dreams." Future 'smart city' scenarios are increasingly drawn from a rich repository of media imaginaries, ranging from Bladerunner-like cityscapes – driverless cars, drone-operated delivery services and hybrid AI machines – to fully-automated communism which promises abundant resources for all and low demands on workers.

Sideways to this discourse, there are plenty of academic research centres courting high-tech industry for funding or partnerships for future venues; and, indeed, high-tech industry courting academic research centres in search of validation for their new technological 'solutions'. Their hope is to reach new market shares in the 'smart city' innovation, this always up-and-coming venue promising an incumbent technological and consumer-ready future. Occasionally, either part meets downtown in one of the many 'smart city' events, these resembling more TED-EX talks than planning sessions. In order to catch up with each other's venues, twitter accounts keep networking and circulating comments and validations, mostly through themselves and through the technology-focused digital magazines they read. These vignettes are snapshots of how the epistemic communities are constructed and work, and how a coalition of advocates advice, suggest, and demand everything 'smart city': This is how a discourse is built and maintained, gets traction and travels in order to become policy.

The future these imaginaries envision is not, however, a common narrative. As Coleman and Tutton (2017, p. 444) question, "Who is the 'we' that becomes affiliated with dystopian or utopian visions of the future? Are different versions of the future possible or desirable for everyone?" I would invite readers to move beyond the dominant framing reproduced by the 'smart city' advocacy coalitions and, rather, ask: Is the adjective 'smart' still appropriate? And, can it be of any use for radical urbanism? Is it rather the case to get going "Beyond the smart city today," as the title of a recent conference call in Rotterdam suggested?

The theoretic grounding of the 'right to the smart city' (Kitchin, Cardullo, et al., 2019) underpins the ethos of this book. This notion takes seriously 'the urban' as the sociocultural dimension and geographical scale in which people live and work. The notion of 'smartness' that emerges from this discussion concerns the complexities and wickedness of city living, fostering an idea of 'the urban' which is very different from the knowable, programmable, and thus linear processes postulated by urban science. The 'right to the city' consists of the mobilisation and full recognition of citizens' use values: "the right of all city dwellers to fully enjoy urban life with all of its services and advantages – the right to habitation – as well as taking direct part in the management of cities – the right to participation" (Fernandes, 2007, p. 208).

The 'right to the smart city', then, is a rallying cry for transformative political mobilization to create a more emancipatory and empowering 'smart city' (Kitchin, Cardullo, et al., 2019). Two ideals transpire from researching the 'smart city' with urban commons in mind. The first one is a Beta City in continual evolution – where stewardship is compelling and hinders the expansion of the commons in the long term (Corsin Jimenez, 2014; de Lange & de Waal, 2013; Leontidou, 2015). We can call this ideal, 'smart commons', which involves open data, ownership of own data, Lo-Fi citizen science projects, and hacker spaces. Despite great potentialities and forms of horizontal organisation, however, I suggest this version of the commons risks becoming instrumental to the neoliberal logics of city growth for the few. The second ideal is rooted in the 'right to the city' – that is, within rights-based and people-focused forms of urban regeneration. These include human rights to decent shelters, ownership of critical infrastructures, forms of remuneration for the data provided, and forms of governance linked to political actions which holds city or state into account (Joss et al., 2017; McFarlane & Söderström, 2017; McLaren & Agyeman, 2015).

This is the right of the excluded, the distressed, and the alienated to demand and receive the material (e.g., a living wage, shelter) and nonmaterial (e.g., recognition, respect, dignity) necessities of life (Marcuse, 2009): it is, therefore, a project of radical democracy and of redistributive justice, which creates cities that are not rooted in and driven by capitalism. It is a right to social justice, which includes but far exceeds the right to individual justice. In this sense it is a common right, not an individual right, and exceeds individual liberty (Harvey, 2008). Thus, I would rather suggest to think in terms of an 'intelligent city' based on collective rights and entitlement which are deeply political and ethical; and an ideal of democratic governance which is grounded on the deliberative power of citizens. In few words, the 'intelligent city' is alternative to the current neoliberal one.

The book has shown some possibilities in turning the tide on neoliberal smart urbanism towards a more democratic and evolving municipalism. Different aspects of this 'return to the public' have been highlighted: from the data and network infrastructure that delivers the Internet and the myriad services associated, to locally sourced digital platforms working on cooperative principles. Thus, there is the need to highlight the politics and ethics of a city *alternative* to the 'smart city', which I call for convenience an 'intelligent city'. Such a configuration entails notions of citizenship that are thoroughly political, reconfigured along collective civil and social rights and entitlements, active and 'meaningful' participation, and redistribution of socially accrued benefits for the common good. In an 'intelligent city', governing bodies and social movements would deliver their municipally owned Internet infrastructure that works for the common good, a *useful* infrastructure for 'the many', a fair and empowering Internet of People rebuilt thanks to the co-operation among small-scale and community network operators, governed through low-cost and open-access network infrastructure, and that operates for the public–community co-production of urban policies. The 'intelligent city' envisioned here, however, is an epitome that works only if we take into consideration the different meanings of the adjective 'smart' and the different cultural constructions attached to each of them.

Notes

1 e.g. www.theguardian.com/world/2020/mar/06/more-scary-than-coronavirus-south-koreas-health-alerts-expose-private-lives
2 www.freepress.net/news/press-releases/free-press-calls-major-internet-providers-waive-broadband-bills-risk-people-and
3 www.freepress.net/our-response/expert-analysis/insights-opinions/racial-digital-divide-persists
4 www.warner.senate.gov/public/_cache/files/b/5/b5ab71aa-ec48-4aa1-9a38-0ce0ddf061 85/8532EB9A7C2E638DD4A099AD23F5478C.3.17.20-google-letter.pdf
5 http://www.youtube.com/watch?v=CAiGKLYsuGY
6 https://www.isinnova.it/easy-covid19-eng/
7 www.dream-smart.city

References

Aalbers, M. B., & Gibb, K. (2014). Housing and the right to the city: Introduction to the special issue. *International Journal of Housing Policy*, *14*(3), 207–213. https://doi.org/10.1080/14616718.2014.936179

Apostol, I., & Antoniadis, P. (2020). Central urban space as a hybrid common infrastructure. *The Journal of Peer Production*, (14).

Arnstein, S. R. (1969). A ladder of citizen participation. *Journal of the American Institute of Planners*, *35*(4), 216–224. https://doi.org/10.1080/01944366908977225

Attoh, K. A. (2011). What kind of right is the right to the city? *Progress in Human Geography*, *35*(5), 669–684. https://doi.org/10.1177/0309132510394706

Baig, R., Roca, R., Navarro, L., & Freitag, F. (2015). guifi.net: A network infrastructure commons. *Proceedings of the Seventh International Conference on Information and Communication Technologies and Development – ICTD '15*, 1–4. https://doi.org/10.1145/2737856.2737900

Bakıcı, T., Almirall, E., & Wareham, J. (2013). A smart city initiative: The case of Barcelona. *Journal of the Knowledge Economy*, *4*(2), 135–148. https://doi.org/10.1007/s13132-012-0084-9

Banerjee, S. B. (2012). The ethics of corporate social responsibility. In *Management ethics* (pp. 69–90). New York and London: Routledge.

Bar, F., & Galperin, H. (2004). Geeks, bureaucrats and cowboys: Deploying internet infrastructure, the wireless way. In M. Castells (Ed.), *The network society: A cross-cultural perspective* (pp. 269–287). Cheltenham: Edward Elgar.

Barandiaran, X., Calleja, A., Monterde, A., Aragón, P., Linares, J., Romero, C., & Pereira, A. (2017). Decidim: Redes políticas y tecnopolíticas para la democracia participativa. *RECERCA. Revista de Pensament y Anàlisi*, (21), 137–150.

Batty, M. (2017). *The new science of cities*. Cambridge, MA: The MIT Press.

Benjamin, W. (1940). *The arcades project*. Cambridge, MA: Belknap Press.

Birkinbine, B. J. (2018). Commons praxis: Toward a critical political economy of the digital commons. *TripleC: Communication, Capitalism & Critique. Open Access Journal for a Global Sustainable Information Society*, *16*(1), 290–305.

Bonnett, A. (2006). The nostalgias of situationist subversion. *Theory, Culture & Society*, *23*(5), 23–48. https://doi.org/10.1177/0263276406067096

Bookchin, D. (2017, July 21). Radical municipalism: The future we deserve. Retrieved November 13, 2017, from ROAR Magazine website: https://roarmag.org/magazine/debbie-bookchin-municipalism-rebel-cities/

Breathnach, P. (2010). From spatial keynesianism to post-fordist neoliberalism: Emerging contradictions in the spatiality of the Irish state. *Antipode*, *42*(5), 1180–1199.

Brenner, N., Marcuse, P., & Mayer, M. (2009). Cities for people, not for profit. *City*, *13*(2), 176–184. https://doi.org/10.1080/13604810903020548

Brenner, N., Peck, J., & Theodore, N. (2010). Variegated neoliberalization: Geographies, modalities, pathways. *Global Networks*, *10*(2), 182–222. https://doi.org/10.1111/j.1471-0374.2009.00277.x

Brenner, N., & Schmid, C. (2015). Towards a new epistemology of the urban? *City*, *19*(2–3), 151–182. https://doi.org/10.1080/13604813.2015.1014712

Brenner, N., & Theodore, N. (2002). Cities and the geographies of "actually existing neoliberalism". *Antipode*, *34*(3), 349–379.

Bresnihan, P., & Byrne, M. (2015). Escape into the city: Everyday practices of commoning and the production of urban space in Dublin. *Antipode*, *47*(1), 36–55.

Brown, W. (2016). Sacrificial citizenship: Neoliberalism, human capital, and austerity politics. *Constellations*, *23*(1), 3–14. https://doi.org/10.1111/1467-8675.12166

Brownill, S., & O'Hara, G. (2015). From planning to opportunism? Re-examining the creation of the London Docklands development corporation. *Planning Perspectives*, *30*(4), 537–570. https://doi.org/10.1080/02665433.2014.989894

Caprotti, F. (2014). Critical research on eco-cities? A walk through the Sino-Singapore Tianjin Eco-city, China. *Cities*, *36*, 10–17.

Caragliu, A., Del Bo, C., & Nijkamp, P. (2011). Smart cities in Europe. *Journal of Urban Technology*, *18*(2), 65–82.

Cardullo, P. (2014). Sniffing the city: Issues of sousveillance in inner city London. *Visual Studies*, *29*(3), 285–293. https://doi.org/10.1080/1472586X.2014.941550

Cardullo, P. (2017). Gentrification in the mesh? *City*, *21*(3–4), 405–419. https://doi.org/10.1080/13604813.2017.1325236

Cardullo, P. (2019). Smart approach to the commons? A case for a public internet infrastructure. In R. Kitchin, P. Cardullo, & C. Di Feliciantonio (Eds.), *The right to the smart city*. Retrieved from https://osf.io/preprints/socarxiv/u8dk2/

Cardullo, P., & Kitchin, R. (2019a). Being a "citizen" in the smart city: Up and down the scaffold of smart citizen participation in Dublin, Ireland. *GeoJournal*, *84*(1), 1–13. https://doi.org/10.1007/s10708-018-9845-8

Cardullo, P., & Kitchin, R. (2019b). Smart urbanism and smart citizenship: The neoliberal logic of "citizen-focused" smart cities in Europe. *Environment and Planning C: Politics and Space*, *37*(5), 813–830. https://doi.org/10.1177/0263774X18806508

Cardullo, P., Kitchin, R., & Di Feliciantonio, C. (2018). Living labs and vacancy in the neoliberal city. *Cities*, *73*, 44–50. https://doi.org/10.1016/j.cities.2017.10.008

Cardullo, P., Kitchin, R., & Di Feliciantonio, C. (Eds.). (2019). *The right to the smart city*. Bingley, UK: Emerald Publishing.

Cardullo, P., & Ribera-Fumaz, R. (forthcoming). *The journey of policy: Barcelona "smart city" reloaded*. Telematics and Informatics, Smart Urbanism Special issue.

Cardullo, P., & Roio (Jaromil), D. (2020). Mesh networking. In *Wiley Blackwell encyclopedia of sociology* (2nd ed.). Retrieved from https://onlinelibrary.wiley.com/doi/abs/10.1002/9781405165518.wbeos1547

Castells, M. (1996). *The rise of the network society*. Malden, MA: Blackwell Publishers.

Castelnovo, W., Misuraca, G., & Savoldelli, A. (2015). Citizen's engagement and value co-production in smart and sustainable cities. In *International conference on public policy*, (pp. 1–16). Retrieved from http://www.icpublicpolicy.org/conference/file/reponse/1433973333.pdf

Castree, N. (2006). From neoliberalism to neoliberalisation: Consolations, confusions, and necessary illusions. *Environment and Planning A*, *38*(1), 1–6.

Charnock, G., March, H., & Ribera-Fumaz, R. (2019). From smart to rebel city? Worlding, provincialising and the Barcelona model. *Urban Studies*, 0042098019872119. https://doi.org/10.1177/0042098019872119

Cheshire, J. (2012) 'Lives on the line: mapping life expectancy along the London Tube network', Environment and Planning A, *44*(7), pp. 1525–1528. doi: 10.1068/a45341.

Clark, D. (2018, December 31). What is 5g? here's what you need to know about the new cellular network. *The New York Times*. Retrieved from www.nytimes.com/2018/12/31/technology/personaltech/5g-what-you-need-to-know.html

Clark, J., & Shelton, T. (2016). Technocratic values and uneven development in the "Smart City." *Metropolitics*. Retrieved from www.metropolitiques.eu/Technocratic-Values-and-Uneven.html

Clarke, J., Newman, J., Smith, N., Vidler, E., & Westmarland, L. (2007). *Creating citizen-consumers: Changing publics and changing public services*. London: SAGE Publications Ltd.

Cohen, N. (2017, December 20). Red hook's cutting-edge wireless network. *The New York Times*. Retrieved from www.nytimes.com/2014/08/24/nyregion/red-hooks-cutting-edge-wireless-network.html

Coleman, R., & Tutton, R. (2017). Introduction to special issue of sociological review on "Futures in Question." *The Sociological Review*, *65*(3), 440–447. https://doi.org/10.1111/1467-954X.12448

Coletta, C., Heaphy, L., & Kitchin, R. (2018). From the accidental to articulated smart city: The creation and work of "Smart Dublin." *European Urban and Regional Studies*, 096977641878521. https://doi.org/10.1177/0969776418785214

Coletta, C., & Kitchin, R. (2016). *Algorithmic governance: Regulating the "heartbeat" of a city using the internet of things*. Retrieved from http://eprints.maynoothuniversity.ie/7622/

Corsin Jimenez, A. (2014). The right to infrastructure: A prototype for open source urbanism. *Environment and Planning D: Society and Space*, *32*(2), 342–362. https://doi.org/10.1068/d13077p

Couldry, N. (2018, October 25). The price of connection. Retrieved November 22, 2018, from HIIG website: www.hiig.de/en/the-price-of-connection/

Cowley, R., Joss, S., & Dayot, Y. (2017). The smart city and its publics: Insights from across six UK cities. *Urban Research & Practice*, *11*(1), 1–25. https://doi.org/10.1080/17535069.2017.1293150

Cugurullo, F. (2018). Exposing smart cities and eco-cities: Frankenstein urbanism and the sustainability challenges of the experimental city. *Environment and Planning A: Economy and Space*, *50*(1), 73–92. https://doi.org/10.1177/0308518X17738535

da Silveira Arruda, N., & Yances, H. (2016). MAP Cartagena: metodología para el mapeo de asentamientos precarios usando OpenStreetMap. *Revista Cartográfica*, (93), 97–116.

Datta, A. (2018). The digital turn in postcolonial urbanism: Smart citizenship in the making of India's 100 smart cities. *Transactions of the Institute of British Geographers*, *43*(3), 325–524. https://doi.org/10.1111/tran.12225

Davidson, C., & Santorelli, M. (2015). *Understanding the debate over government-owned broadband networks*. Retrieved from https://bit.ly/2MDk1tc

Davies, W. (2015). *The happiness industry: How the government and big business sold us well-being*. London: Verso.

De Angelis, M. (2001). Marx and primitive accumulation: The continuous character of capital's "enclosures." *The Commoner*, *2*. Retrieved from http://thecommoner.org/

De Angelis, M. (2017). *Omnia sunt communia: On the commons and the transformation to postcapitalism*. London: Zed Books.

De Angelis, M., & Stavrides, S. (2010, June). On the commons: A public interview with massimo De Angelis and Stavros Stavrides. *An Architektur, e-Flux, Journal, 17*. Retrieved from www.e-flux.com/journal/17/67351/on-the-commons-a-public-interview-with-massimo-de-angelis-and-stavros-stavrides/

de Lange, M., & de Waal, M. (2013). Owning the city: New media and citizen engagement in urban design. *First Monday, 18*(11). Retrieved from http://firstmonday.org/ojs/index.php/fm/article/view/4954

de Lange, M., & de Waal, M. (2016). *Hacking Buiksloterham*. Retrieved from http://the-hackablecity.nl/

Deleuze, G. (2002). Postscript to control society. In T. Y. Levin & U. Frohne (Eds.), *Ctrl [space]: Rhetorics of surveillance from Bentham to Big Brother*. Karlsruhe, Germany: ZKM Center for Art and Media.

Della Porta, D., Fernández, J., Kouki, H., & Mosca, L. (2017). *Movement parties against austerity*. Cambridge, UK and Malden, MA: Polity Press.

de Waal, M. (2014). *The city as interface: How digital media are changing the city*. Rotterdam: nai010 publishers.

Diggelmann, O., & Cleis, M. N. (2014). How the right to privacy became a human right. *Human Rights Law Review, 14*(3), 441–458.

D'Ignazio, C., & Klein, L. F. (2020). *Data feminism*. Cambridge, MA: The MIT Press.

Di Feliciantonio, C. (2019). Against the romance of the smart community: The case of Milano 4 You. In P. Cardullo, R. Kitchin, & C. Di Feliciantonio (Eds.), *The right to the smart city*. Bingley, UK: Emerald Publishing.

Ding, J. (2018). *Deciphering China's AI dream*. Future of Humanity Institute Technical Report.

Dodge, M., & Kitchin, R. (2001). *Atlas of cyberspace* (1st ed.). Harlow: Addison-Wesley.

Dodge, M., & Kitchin, R. (2003). *Mapping cyberspace*. London; New York: Routledge.

Dodge, M., & Kitchin, R. (2013). Crowdsourced cartography: Mapping experience and knowledge. *Environment and Planning A, 45*(1), 19–36. https://doi.org/10.1068/a44484

Douglas, H. (1963). *The underground story*. London: R. Hale.

Dutilleul, B., Birrer, F. A., & Mensink, W. (2010). Unpacking European living labs: Analysing innovation's social dimensions. *Central European Journal of Public Policy, 4*(1), 60–85.

Engelbert, J., van Zoonen, L., & Hirzalla, F. (2019, May). Excluding citizens from the European smart city: The discourse practices of pursuing and granting smartness. *Technological Forecasting and Social Change, 142*, 347–353. https://doi.org/10.1016/j.techfore.2018.08.020

EPB (2015). EPB to offer discounted internet for low-income families. Retrieved March 27, 2018, from EPB | Powering Chattanooga website: https://epb.com/about-epb/news/articles/8

Eubanks, V. (2018). *Automating inequality: How high-tech tools profile, police, and punish the poor*. New York, NY: St. Martin's Press.

European Commission (2016). *Horizon 2020 work programme 2016–2017* (Cross-Cutting Activities (Focus Areas) No. 17).

Evans, J. P. M., & Karvonen, A. (2014). "Give Me a Laboratory and I Will Lower Your Carbon Footprint!" – Urban laboratories and the governance of low-carbon futures in Manchester. *International Journal of Urban and Regional Research, 38*(2), 413–430. https://doi.org/10.1111/1468-2427.12077

Evans, J. P. M., Karvonen, A., & Raven, R. (Eds.). (2016). *The experimental city*. London and New York, NY: Routledge, Taylor & Francis Group.

Featherstone, M. (2009). Ubiquitous media. *Theory, Culture & Society, 26*(2–3), 1.

Fernandes, E. (2007). Constructing the "Right to the City" in Brazil. *Social & Legal Studies, 16*(2), 201–219. https://doi.org/10.1177/0964663907076529

Flanagan, K., Connolly, M., Brown, S., & Cox, M. (2019). Report on the impacts of Airbnb in the republic of Ireland. Retrieved from Irish Housing Network; Inside Airbnb website: https://borderinggeographies.files.wordpress.com/2020/01/report-on-the-impacts-of-airbnb-in-ireland.pdf

Florida, R. L. (2003). *The rise of the creative class: And how it's transforming work, leisure, community and everyday life*. North Melbourne, VIC: Pluto Press.

Floridi, L. (2019). Establishing the rules for building trustworthy AI. *Nature Machine Intelligence, 1*. https://doi.org/10.1038/s42256-019-0055-y

Foth, M. (2016). Why we should design smart cities for getting lost. Retrieved February 3, 2017, from The Conversation website: http://theconversation.com/why-we-should-design-smart-cities-for-getting-lost-56492

Foth, M. (2017). The software-sorted city: Big data and algorithms. In *Digital Cities Workshop*. Retrieved from https://dc10web.wordpress.com

Foth, M., Klaebe, H., Adkins, B., & Hearn, G. (2009). Embedding an ecology notion in the social production of urban space. In M. Foth (Ed.), *Handbook of research on urban informatics: The practice and promise of the real-time city* (pp. 179–194). Hershey, PA: Information Science Reference.

Foucault, M. (2008). *The birth of biopolitics: Lectures at the Collège de France, 1978–1979* (A. I. Davidson & G. Burchell, Eds.). Basingstoke, Hampshire; New York, NY: Palgrave Macmillan.

Fuchs, C. (2018). Industry 4.0: The digital German ideology. *TripleC: Communication, Capitalism & Critique. Open Access Journal for a Global Sustainable Information Society, 16*(1), 280–289. https://doi.org/10.31269/triplec.v16i1.1010

Fuller, M. (Ed.). (2017). *How to be a geek: Essays on the culture of software*. Cambridge, UK and Malden, MA: Polity Press.

Gabrys, J. (2011). *Digital rubbish: A natural history of electronics*. Ann Arbor: University of Michigan Press.

Gabrys, J. (2014). Programming environments: Environmentality and citizen sensing in the smart city. *Environment and Planning D: Society and Space, 32*(1), 30–48. https://doi.org/10.1068/d16812

Gabrys, J., Pritchard, H., & Barratt, B. (2016). Just good enough data: Figuring data citizenships through air pollution sensing and data stories. *Big Data & Society, 3*(2), 2053951716679677. https://doi.org/10.1177/2053951716679677

Galdon, G. (2017, April 25). Technological sovereignty? Democracy, data and governance in the digital era. Retrieved January 31, 2020, from CCCB LAB website: http://lab.cccb.org/en/technological-sovereignty-democracy-data-and-governance-in-the-digital-era/

Gerbaudo, P. (2019). The platform party: The transformation of political organisation in the era of big data. In D. Chandler & C. Fuchs (Eds.), *Digital objects, digital subjects: Interdisciplinary perspectives on capitalism, labour and politics in the age of big data* (pp. 187–198). https://doi.org/10.16997/book29.p

Gonzalez, L. (2018, March 13). Grassroots group forming to take action on broadband in Massachusetts city. Retrieved March 29, 2018, from Institute for Local Self-Reliance website: https://ilsr.org/grassroots-group-forming-to-take-action-on-broadband-in-massachusetts-city/

Graham, S., & Aurigi, A. (1997). Urbanising cyberspace?: The nature and potential of the virtual cities movement. *City*, *2*(7), 18–39. https://doi.org/10.1080/13604819708900051

Graham, S., & Marvin, S. (1996). *Telecommunications and the city: Electronic spaces, urban places*. New York and London: Routledge.

Graham, S., & Marvin, S. (2001). *Splintering urbanism: Networked infrastructures, technological mobilities and the urban condition*. London: Routledge.

Greenfield, A. (2013). *Against the smart city* (1.3 ed.). New York, NY: Do projects.

Greenfield, A. (2017). *Radical technologies: The design of everyday life*. Retrieved from www.versobooks.com/books/2453-radical-technologies

Han, B.-C. (2017). *Psychopolitics: Neoliberalism and new technologies of power*. Brooklyn: Verso Books.

Hardt, M., & Negri, A. (2009). *Commonwealth*. Cambridge, MA: Belknap Press of Harvard University Press.

Harvey, D. (1978). The urban process under capitalism: A framework for analysis. *International Journal of Urban and Regional Research*, *2*(1–4), 101–131.

Harvey, D. (2003). The right to the city. *International Journal of Urban and Regional Research*, *27*(4), 939–941.

Harvey, D. (2008). The right to the city. *New Left Review*, *53*. Retrieved from https://newleftreview.org/issues/II53/articles/david-harvey-the-right-to-the-city

Harvey, D. (2011). The future of the commons. *Radical History Review*, *2011*(109), 101–107. https://doi.org/10.1215/01636545-2010-017

Harvey, D., Negri, A., & Hardt, M. (2009, November). Commonwealth: An exchange. *Artforum*. Retrieved from http://libcom.org/library/commonwealth-exchange

Hayek, F. A. (2011). *The constitution of liberty: The definitive edition*. New York and London: Routledge.

Hayles, N. K. (2009). RFID: Human agency and meaning in information-intensive environments. *Theory, Culture & Society*, *26*(2–3), 47–72. https://doi.org/10.1177/0263276409103107

Heaphy, L., & Pétercsák, R. (2016, September 5). *Building smart city partnerships in the "Silicon Docks."* Paper presented at the Creating Smart Cities Workshop. Presented at the Maynooth University, Ireland. Retrieved from http://bit.ly/2mjmyfd

Helbich, M., Jokar Arsanjani, J., & Leitner, M. (Eds.). (2015). *Computational approaches for urban environments*. https://doi.org/10.1007/978-3-319-11469-9

Hemment, D., & Townsend, A. (Eds.). (2013). *Smart citizens*. Retrieved from https://waag.org/sites/waag/files/media/publicaties/smartcitizens.pdf

Hill, D. (2013). Essay: On the smart city; or, a "manifesto" for smart citizens instead. Retrieved October 26, 2017, from Cityofsound website: www.cityofsound.com/blog/2013/02/on-the-smart-city-a-call-for-smart-citizens-instead.html

Hindess, B. (2002). Neo-liberal citizenship. *Citizenship Studies*, *6*(2), 127–143. https://doi.org/10.1080/13621020220142932

Hoepman, J.-H. (2019). *Privacy design strategies*. CC 4.0.

Hollands, R. G. (2008). Will the real smart city please stand up? Intelligent, progressive or entrepreneurial? *City*, *12*(3), 303–320.

Hollands, R. G. (2015). Critical interventions into the corporate smart city. *Cambridge Journal of Regions, Economy and Society*, *8*(1), 61–77. https://doi.org/10.1093/cjres/rsu011

Hu, T.-H. (2015). *A prehistory of the cloud*. Cambridge, MA: The MIT Press.

Huron, A. (2015). Working with strangers in saturated space: Reclaiming and maintaining the urban commons: The urban commons. *Antipode*, *47*(4), 963–979. https://doi.org/10.1111/anti.12141

Iaione, C. (2012). City as a commons. *Second Thematic Conference of the IASC on "Design and Dynamics of Institutions for Collective Action": A Tribute to Prof. Elinor Ostrom, 29.* Retrieved from http://papers.ssrn.com/sol3/papers.cfm?abstract_id=2589640

Illich, I. (1982). *Silence is a commons.* Retrieved from www.preservenet.com/theory/Illich/Silence.html

Ishida, T., & Isbister, K. (2000). *Digital cities: Technologies, experiences, and future perspectives.* Berlin: Springer.

Isin, E. F. (Ed.). (2000). *Democracy, citizenship, and the global city.* London and New York, NY: Routledge.

Isin, E. F., & Ruppert, E. (2015). *Being digital citizens.* London: Rowman & Littlefield.

Isin, E. F., & Wood, P. K. (1999). *Citizenship and identity.* London, UK: Sage.

Joss, S., Cook, M., & Dayot, Y. (2017). Smart cities: Towards a new citizenship regime? A discourse analysis of the British smart city standard. *Journal of Urban Technology,* 1–21. https://doi.org/10.1080/10630732.2017.1336027

Karvonen, A., Cugurullo, F., & Caprotti, F. (2018). *Inside smart cities: Place, politics and urban innovation.* New York and London: Routledge.

Keith, M., & Pile, S. (1993). *Place and the politics of identity.* New York and London: Routledge.

Kenney, M., & Zysman, J. (2015). Choosing a future in the platform economy: The implications and consequences of digital platforms. *Kauffman Foundation New Entrepreneurial Growth Conference, 156160.*

Kim, Y.-L. (2018). Seoul's Wi-Fi hotspots: Wi-Fi access points as an indicator of urban vitality. *Computers, Environment and Urban Systems, 72,* 13–24. https://doi.org/10.1016/j.compenvurbsys.2018.06.004

Kitchin, R. (2014a). *The data revolution: Big data, open data, data infrastructures & their consequences.* London: Sage Publications.

Kitchin, R. (2014b). The real-time city? Big data and smart urbanism. *GeoJournal, 79*(1), 1–14. https://doi.org/10.1007/s10708-013-9516-8

Kitchin, R. (2015). Making sense of smart cities: Addressing present shortcomings. *Cambridge Journal of Regions, Economy and Society, 8*(1), 131–136. https://doi.org/10.1093/cjres/rsu027

Kitchin, R. (2016a). *Reframing, reimagining and remaking smart cities.* The Programmable City Working Paper 20. https://doi.org/10.17605/OSF.IO/CYJHG

Kitchin, R. (2016b). The ethics of smart cities and urban science. *Philosophical Transactions of the Royal Society A, 374*(2083), 20160115. https://doi.org/10.1098/rsta.2016.0115

Kitchin, R. (2017). *Urban science: A short primer.* The Programmable City Working Paper 23.

Kitchin, R. (2019a). *The ethics of smart cities.* Retrieved from www.rte.ie/brainstorm/2019/0425/1045602-the-ethics-of-smart-cities/

Kitchin, R. (2019b). The timescape of smart cities. *Annals of the American Association of Geographers, 109*(3), 775–790. https://doi.org/10.1080/24694452.2018.1497475

Kitchin, R. (2020). Civil liberties or public health, or civil liberties and public health? Using surveillance technologies to tackle the spread of COVID-19. *Space and Polity, 0*(0), 1–20. https://doi.org/10.1080/13562576.2020.1770587

Kitchin, R., Cardullo, P., & Di Feliciantonio, C. (2019). Citizenship, justice and the right to the smart city. *Right to the Smart City.* Retrieved from https://doi.org/10.31235/osf.io/b8aq5

Kitchin, R., Coletta, C., Evans, L., Heaphy, L., & MacDonncha, D. (2017). Smart cities, epistemic communities, advocacy coalitions and the "last mile" problem. *it – Information Technology, 59*(6), 275–284. https://doi.org/10.1515/itit-2017-0004

Kitchin, R., Coletta, C., & McArdle, G. (2017). *Urban informatics, governmentality and the logics of urban control.* Programmable City Working Paper 25. Retrieved from https://osf.io/preprints/socarxiv/27hz8/

Kitchin, R., & Dodge, M. (2011). *Code/Space: Software and everyday life.* Cambridge, MA: The MIT Press.

Kitchin, R., & Dodge, M. (2019). The (In)Security of smart cities: Vulnerabilities, risks, mitigation, and prevention. *Journal of Urban Technology, 26*(2), 47–65. https://doi.org/10.1080/10630732.2017.1408002

Kitchin, R., Graham, M., Mattern, S., & Shaw, J. (Eds.). (2019). *How to run a city like Amazon, and other fables.* London: Meatspace Press.

Kitchin, R., Maalsen, S., & McArdle, G. (2016). The praxis and politics of building urban dashboards. *Geoforum, 77*, 93–101.

Kitchin, R., O'Callaghan, C., Boyle, M., Gleeson, J., & Keaveney, K. (2012). Placing neoliberalism: The rise and fall of Ireland's Celtic Tiger. *Environment and Planning A, 44*(6), 1302–1326. https://doi.org/10.1068/a44349

Kitheka, B. M., Baldwin, E. D., White, D. L., & Harding, D. N. (2016). A different "we" in urban sustainability: How the city of Chattanooga, TN, community defined their own sustainability path. *International Journal of Tourism Cities, 2*(3), 185–205. https://doi.org/10.1108/IJTC-07-2015-0017

Koebler, J. (2016, January 25). How a DIY Network plans to subvert Time Warner cable's NYC internet monopoly. Retrieved December 31, 2018, from Motherboard website: https://motherboard.vice.com/en_us/article/gv5qb4/how-a-diy-network-plans-to-subvert-time-warner-cables-nyc-internet-monopoly

Larkin, B. (2013). The politics and poetics of infrastructure. *Annual Review of Anthropology, 42*(1), 327–343. https://doi.org/10.1146/annurev-anthro-092412-155522

Lash, S. (2002). *Critique of information.* London and Thousand Oaks, CA: Sage Publications.

Lawton, P., & Punch, M. (2014). Urban governance and the "European City": Ideals and realities in Dublin, Ireland. *International Journal of Urban and Regional Research, 38*(3), 864–885. https://doi.org/10.1111/1468-2427.12152

Lazonick, W., & Mazzucato, M. (2013). The risk-reward nexus in the innovation-inequality relationship: Who takes the risks? Who gets the rewards? *Industrial and Corporate Change, 22*(4), 1093–1128.

Lefebvre, H. (1996). The right to the city. In E. Kofman & E. Lebas (Eds.), *Writings on cities.* Cambridge, MA: Blackwell.

Leontidou, L. (2015). "Smart Cities" of the debt crisis: Grassroots creativity in Mediterranean Europe. *Επιθεώρηση Κοινωνικών Ερευνών, 144*(144), 69–101. https://doi.org/10.12681/grsr.8626

Lewis-Kraus, G. (2016, November 3). The battle to bring broadband to New York's public housing. *Wired.* Retrieved from www.wired.com/2016/11/bringing-internet-to-new-york-public-housing/

Li, R., Clemm, A., Chunduri, U., Dong, L., & Makhijani, K. (2018). A new framework and protocol for future networking applications. In *Proceedings of the 2018 workshop on networking for emerging applications and technologies - NEAT '18* (pp. 21–26). Budapest: ACM Press.

Light, J. S. (2002). Urban security from warfare to welfare. *International Journal of Urban and Regional Research, 26*(3), 607–613. https://doi.org/10.1111/1468-2427.00403

Ljungberg, J. (2000). Open source movements as a model for organising. *European Journal of Information Systems, 9*(4), 208–216. https://doi.org/10.1057/palgrave.ejis.3000373

Luque-Ayala, A., & Marvin, S. (2016). The maintenance of urban circulation: An operational logic of infrastructural control. *Environment and Planning D: Society and Space*, *34*(2), 191–208. https://doi.org/10.1177/0263775815611422

Lyon, D. (1994). *The electronic eye: The rise of surveillance society*. Cambridge, MA: Polity Press.

Mabud, R., & Seitz-Brown, M. (2017). *Wired: Connecting equity to a universal broadband strategy*. The Roosevelt Institute. Retrieved from http://rooseveltinstitute.org/wp-content/uploads/2017/09/Wired_Roosevelt-Institute.pdf

MacLaran, A., & Kelly, S. (2014). Irish neoliberalism and neoliberal urban policy. In *Neoliberal urban policy and the transformation of the city* (pp. 20–36). https://doi.org/10.1057/9781137377050_2

MacLeod, G. (2011). Urban politics reconsidered: Growth machine to post-democratic city? *Urban Studies*, *48*(12), 2629–2660. https://doi.org/10.1177/0042098011415715

Madden, D. J., & Marcuse, P. (2016). *Intro | In defense of housing: The politics of crisis*. London: Verso.

Malmgren, E. (2017, December). The new sewer socialists. Retrieved January 23, 2020, from Logic Magazine website: https://logicmag.io/justice/the-new-sewer-socialists/

March, H. (2018). The Smart City and other ICT-led techno-imaginaries: Any room for dialogue with Degrowth? *Journal of Cleaner Production*, *197*, 1694–1703. https://doi.org/10.1016/j.jclepro.2016.09.154

March, H., & Ribera-Fumaz, R. (2016). Smart contradictions: The politics of making Barcelona a self-sufficient city. *European Urban and Regional Studies*, *23*(4), 816–830. https://doi.org/10.1177/0969776414554488

Marcuse, P. (1985). Gentrification, abandonment, and displacement: Connections, causes, and policy responses in New York City. *Washington University Journal of Urban and Contemporary Law*, *28*, 195–240.

Marcuse, P. (2009). From critical urban theory to the right to the city. *City*, *13*(2), 185–197. https://doi.org/10.1080/13604810902982177

Marshall, T. H. (1992). *Citizenship and social class* (T. Bottomore, Ed.). London: Pluto Press.

Massey, D. B. (2007). *World city*. Cambridge, MA: Polity Press.

Mattern, S. (2016). Instrumental city: The view from Hudson Yards, circa 2019. *Places Journal*. https://doi.org/10.22269/160426

Mattern, S. (2017). A city is not a computer. *Places Journal*. https://doi.org/10.22269/170207

Mattern, S. (2019). Networked dream worlds. Retrieved September 23, 2019, from Real Life website: https://reallifemag.com/networked-dream-worlds/

Mazzucato, M. (2011). The entrepreneurial state. *Soundings*, *49*(12), 131–142. https://doi.org/10.3898/136266211798411183

Mazzucato, M. (2018a). *The value of everything: Makers and takers in the global economy*. London: Allen Lane.

Mazzucato, M. (2018b, February 21). Mission-oriented research & innovation in the European Union: A problem-solving approach to fuel innovation-led growth. [Website]. Retrieved March 16, 2018, from https://publications.europa.eu/en/publication-detail/-/publication/5b2811d1-16be-11e8-9253-01aa75ed71a1/language-en

McCann, B. (2014). *A review of SCATS operation and deployment in Dublin*. Paper presented at the 19th JCT Traffic Signal Symposium & Exhibition. Retrieved from http://bit.ly/2pdLCs5

McCann, E., & Ward, K. (2011). Urban assemblages: Territories, relations, practices, and power. In E. McCann & K. Ward (Eds.), *Mobile urbanism: Cities and policymaking in the global age*. Minneapolis: University of Minnesota Press.

McCann, E., & Ward, K. (2013). A multi-disciplinary approach to policy transfer research: Geographies, assemblages, mobilities and mutations. *Policy Studies*, *34*(1), 2–18. https://doi.org/10.1080/01442872.2012.748563

McDermott, M. (2019). *New York City internet health report*. Retrieved from Mozilla website: https://foundation.mozilla.org/en/opportunity/new-york-city-internet-health-report/

McFarlane, C., & Söderström, O. (2017, June). On alternative smart cities: From a technology-intensive to a knowledge-intensive smart urbanism. *City*, 1–17. https://doi.org/10.1080/13604813.2017.1327166

McLaren, D., & Agyeman, J. (2015). *Sharing cities: A case for truly smart and sustainable cities*. Cambridge, MA: The MIT Press.

Medosch, A. (2003). *Freie netze*. Hannover: Heise Heinz.

Medosch, A. (2015). Cities of the Sun: Urban revolutions and the network commons. Retrieved February 8, 2019, from The Next Layer website: https://webarchiv.servus.at/thenextlayer.org/node/1358.html

Mengoli, P., & Russo, M. (2017). *A hybrid space to support the regeneration of competences for re-industrialization*. DEMB WORKING PAPER SERIES, Dipartimento di Economia Marco Biagi - Università di Modena e Reggio Emilia.

Metzinger, T. (2019). *Ethics washing made in Europe*. Retrieved from https://www.tagesspiegel.de/politik/eu-guidelines-ethics-washing-made-in-europe/24195496.html

Mezzadra, S., & Neilson, B. (2017). On the multiple frontiers of extraction: Excavating contemporary capitalism. *Cultural Studies*, *31*(2–3), 185–204. https://doi.org/10.1080/09502386.2017.1303425

Mitchell, D. (2003). *The right to the city: Social justice and the fight for public space*. New York, NY: Guilford Press.

Mitchell, W. J. (2007). Intelligent cities. *UOC Papers*, *5*. Retrieved from www.uoc.edu/uocpapers/5/dt/eng/mitchell.pdf

Monnier, J.-M., & Vercellone, C. (2017). *Basic income as primary income*. Université Paris1 Panthéon-Sorbonne (Post-Print and Working Papers). Retrieved from HAL website: https://econpapers.repec.org/paper/halcesptp/hal-01486202.htm

Morozov, E. (2013). *To save everything, click here: Technology, solutionism, and the urge to fix problems that don't exist*. London: Penguin Books.

Morozov, E., & Bria, F. (2018). Rethinking the Smart City. *Rosa Luxemburg Stiftung Nyc*. Retrieved from www.rosalux-nyc.org/rethinking-the-smart-city/

Mosco, V. (2019). The history of the internet under surveillance capitalism. *Science as Culture*, 1–5. https://doi.org/10.1080/09505431.2019.1623191

Mouffe, C. (1999). Deliberative democracy or agonistic pluralism? *Social Research*, *66*(3), 745–758.

Mumford, L. (2009). The highway and the city. In C. Hanks (Ed.), *Technology and values: Essential readings*. Chichester, UK; Malden, MA: Wiley-Blackwell.

Nel·lo, O. (2015). Movimientos urbanos y defensa del patrimonio colectivo en la región metropolitana de Barcelona. *Ciudad y territorio: Estudios territoriales*, (184), 311–327.

Newman, K., & Wyly, E. K. (2006). The right to stay put, revisited: Gentrification and resistance to displacement in New York City. *Urban Studies*, *43*(1), 23–57. https://doi.org/10.1080/00420980500388710

O'Callaghan, C., Feliciantonio, C. D., & Byrne, M. (2017). Governing urban vacancy in post-crash Dublin: Contested property and alternative social projects. *Urban Geography*, 1–24. https://doi.org/10.1080/02723638.2017.1405688

O'Callaghan, C., Kelly, S., Boyle, M., & Kitchin, R. (2015). Topologies and topographies of Ireland's neoliberal crisis. *Space and Polity*, *19*(1), 31–46.

O'Callaghan, C., & Lawton, P. (2016). Temporary solutions? Vacant space policy and strategies for re-use in Dublin. *Irish Geography*, *48*(1), 69–87. https://doi.org/10.2014/igj. v48i1.526

O'Mahony, E., & Rigney, S. (2016). "What's the Story Buddleia?": A public geography of dereliction in Dublin City. *Irish Geography*, *48*(1), 88–99.

O'Neil, C. (2017). *Weapons of math destruction: How big data increases inequality and threatens democracy*. London: Penguin Books. Ebook ISBN 9780553418828.

Ong, A. (2006). Mutations in citizenship. *Theory, Culture & Society*, *23*(2–3), 499–505. https://doi.org/10.1177/0263276406064831

Paasche, T. F. (2013). Coded police territories: "Detective software" investigates *Area*, *45*(3), 314–320. https://doi.org/10.1111/area.12033

Papacharissi, Z. (2010). Privacy as a luxury commodity. *First Monday*, *15*(8). https://doi. org/10.5210/fm.v15i8.3075

Pasquale, F. (2019, April 3). Quantifying love [Text]. Retrieved April 14, 2019, from Boston Review website: http://bostonreview.net/print-issues-politics/frank-pasquale-quantifying-love

Peck, J. (2004). Geography and public policy: Constructions of neoliberalism. *Progress in Human Geography*, *28*(3), 392–405.

Peck, J. (2012). Austerity urbanism: American cities under extreme economy. *City*, *16*(6), 626–655. https://doi.org/10.1080/13604813.2012.734071

Peck, J., Theodore, N., & Brenner, N. (2009). Neoliberal urbanism: Models, moments, mutations. *SAIS Review of International Affairs*, *29*(1), 49–66.

Peck, J., Theodore, N., & Brenner, N. (2013). Neoliberal urbanism redux?: Debates and developments. *International Journal of Urban and Regional Research*, *37*(3), 1091–1099. https://doi.org/10.1111/1468-2427.12066

Peck, J., & Whiteside, H. (2017). Neoliberalizing Detroit. In S. F. Schram & M. Pavlovskaya (Eds.), *Rethinking neoliberalism: Resisting the disciplinary regime* (pp. 179–196). London: Routledge.

Perez, C. (2009). Technological revolutions and techno-economic paradigms. *Cambridge Journal of Economics*, *34*(1), 185–202.

Perng, S.-Y. (2016, September 5). *Creating infrastructure with citizens: Traffic light box artworks in Dublin streets*. Paper presented at the Creating Smart Cities Workshop. Presented at the Maynooth University, Ireland. Retrieved from http://bit.ly/2mjmyfd

Perng, S.-Y. (2017). *Practices and politics of collaborative urban infrastructuring: Traffic light box artworks in Dublin streets*. The Programmable City Working Paper 33. https:// doi.org/10.17605/OSF.IO/2XPQ7

Perng, S.-Y., & Kitchin, R. (2016). Solutions and frictions in civic hacking: Collaboratively designing and building wait time predictions for an immigration office. *Social & Cultural Geography*, 1–20. https://doi.org/10.1080/14649365.2016.1247193

Perng, S.-Y., Kitchin, R., & Evans, L. (2016). Locative media and data-driven computing experiments. *Big Data & Society*, *3*(1), 2053951716652161. https://doi.org/10.1177/2053951716652161

Perng, S.-Y., Kitchin, R., & Mac Donncha, D. (2017). *Hackathons, entrepreneurship and the passionate making of smart cities*. The Programmable City Working Paper 28. Retrieved from osf.io/nu3ec

Picon, A. (2015). *Smart cities: A spatialised intelligence*. Chichester, West Sussex: Wiley.

Plantin, J.-C., & Punathambekar, A. (2018). Digital media infrastructures: Pipes, platforms, and politics. *Media, Culture & Society*, 0163443718818376. https://doi. org/10.1177/0163443718818376

Purcell, M. (2003). Citizenship and the right to the global city: Reimagining the capitalist world order. *International Journal of Urban and Regional Research, 27*(3), 564–590. https://doi.org/10.1111/1468-2427.00467

Ratti, C. (2009, October 2). Digital cities: "Sense-able" urban design. *Wired UK.* Retrieved from www.wired.co.uk/article/digital-cities-sense-able-urban-design

Rhodes, R. A. W. (1996). The new governance: Governing without government. *Political Studies, 44*(4), 652–667. https://doi.org/10.1111/j.1467-9248.1996.tb01747.x

Ribera-Fumaz, R. (2019). Moving from smart citizens to technological sovereignty? In P. Cardullo, R. Kitchin, & C. Di Feliciantonio (Eds.), *The right to the Smart City.* Bingley, UK: Emerald Publishing.

Robinson, J. (2006). *Ordinary cities: Between modernity and development.* London: Routledge.

Rodgers, D., & O'Neill, B. (2012). Infrastructural violence: Introduction to the special issue. *Ethnography, 13*(4), 401–412. https://doi.org/10.1177/1466138111435738

Rossi, U. (2017). *Cities in global capitalism.* Cambridge, UK and Malden, MA: Polity Press.

Roy, A. (2011). Slumdog cities: Rethinking subaltern urbanism. *International Journal of Urban and Regional Research, 35*(2), 223–238. https://doi.org/10.1111/j.1468-2427.2011.01051.x

Rubio-Pueyo, V. (2017). Municipalism in Spain. Retrieved from Rosa Luxemburg Stiftung NYC website: www.rosalux-nyc.org/municipalism-in-spain/

Rushe, D. (2014, August 30). Chattanooga's Gig: How one city's super-fast internet is driving a tech boom. *The Guardian.* Retrieved from www.theguardian.com/world/2014/aug/30/chattanooga-gig-high-speed-internet-tech-boom

Rutland, T. (2010). The financialization of urban redevelopment. *Geography Compass, 4*(8), 1167–1178.

Sadowski, J., & Pasquale, F. (2015). The spectrum of control: A social theory of the smart city. *First Monday, 20*(7). https://doi.org/10.5210/fm.v20i7.5903

Sartori, L. (2015). Alla ricerca della "smart citizenship." *Istituzioni Del Federalismo: Rivista Di Studi Giuridici e Politici,* (4), 927–948.

Schneider, N. (2018, February). Next, the internet: Building a cooperative digital space. *The Cooperative Business Journal.* Retrieved from https://ioo.coop/2018/02/next-the-internet-building-a-cooperative-digital-space/

Scholz, T. (2016). Platform cooperativism. *Rosa Luxemburg Stiftung NYC.* Retrieved from www.rosalux-nyc.org/platform-cooperativism-2/

Scholz, T. (2017). *Uberworked and underpaid.* Retrieved from http://pombo.free.fr/treborscholz.pdf

Seltzer, E., & Mahmoudi, D. (2013). Citizen participation, open innovation, and crowdsourcing: Challenges and opportunities for planning. *CPL Bibliography, 28*(1), 3–18. https://doi.org/10.1177/0885412212469112

Sennett, R. (1992). *The conscience of the eye: The design and social life of cities.* New York, NY: W.W. Norton & Co., by arrangement with Alfred A. Knopf.

Shaw, J., & Graham, M. (2017a). An informational right to the city? Code, content, control, and the urbanization of information. *Antipode, 49*(4), 907–927. https://doi.org/10.1111/anti.12312

Shaw, J., & Graham, M. (2017b). *Our digital rights to the city.* Retrieved from http://meatspacepress.org/#selfprint

Shelton, T., & Lodato, T. (2019). Actually existing smart citizens: Expertise and (non) participation in the making of the smart city. *City,* 1–18.

Shelton, T., Zook, M., & Wiig, A. (2015). The "actually existing smart city." *Cambridge Journal of Regions, Economy and Society, 8*(1), 13–25. https://doi.org/10.1093/cjres/rsu026

Shepard, M. (2011). *Sentient city: Ubiquitous computing, architecture, and the future of urban space.* New York, NY; Cambridge, MA: The MIT Press.

Sheppard, E., Leitner, H., & Maringanti, A. (2013). Provincializing global urbanism: A manifesto. *Urban Geography, 34*(7), 893–900. https://doi.org/10.1080/02723638.2013.807977

Shin, H., Park, S. H., & Sonn, J. W. (2015). The emergence of a multiscalar growth regime and scalar tension: The politics of urban development in Songdo New City, South Korea. *Environment and Planning C: Government and Policy, 33*(6), 1618–1638.

Skeggs, B., & Loveday, V. (2012). Struggles for value: Value practices, injustice, judgment, affect and the idea of class. *The British Journal of Sociology, 63*(3), 472–490.

Skeggs, B., & Yuill, S. (2018). Subjects of value and digital personas: Reshaping the bourgeois subject, unhinging property from personhood. *Subjectivity.* https://doi.org/10.1057/s41286-018-00063-4

Slater, J., & Iles, A. (2009, November). *No room to move: Radical art and the regenerate city.* MetaMute. Retrieved from https://www.metamute.org/editorial/articles/no-room-to-move-radical-art-and-regenerate-city

Slater, T. (2009). Missing Marcuse: On gentrification and displacement. *City, 13*(2), 292–311. https://doi.org/10.1080/13604810902982250

Smart, A. (2018). *Does formalization make a city smarter? Towards post-elitist and posthumanist smart cities.* Paper presented at Smart cities, smart citizens?, Hong Kong City University.

Smith, A. and Pietro Martín, P. (2020) 'Going beyond the smart city? Implementing technopolitical platforms for urban democracy in Madrid and Barcelona', Journal of Urban Technology. Routledge. Available at: https://ictlogy.net/bibliography/reports/projects.php?idp=4186 (Accessed: 23 June 2020).

Söderström, O., Paasche, T., & Klauser, F. (2014). Smart cities as corporate storytelling. *City, 18*(3), 307–320. https://doi.org/10.1080/13604813.2014.906716

Staeheli, L. A. (2011). Political geography: Where's citizenship? *Progress in Human Geography, 35*(3), 393–400. https://doi.org/10.1177/0309132510370671

Stavrides, S. (2016). *Common space: The city as commons.* London: Zed Books Ltd.

Stehle, S., & Kitchin, R. (2020). Real-time and archival data visualisation techniques in city dashboards. *International Journal of Geographical Information Science, 34*(2), 344–366. https://doi.org/10.1080/13658816.2019.1594823

Susser, I. (2017). For or against commoning? *Focaal, 2017*(79), 1–5.

Swyngedouw, E. (2005). Governance innovation and the citizen: The Janus face of governance-beyond-the-state. *Urban Studies, 42*(11), 1991–2006.

Swyngedouw, E. (2011). Interrogating post-democratization: Reclaiming egalitarian political spaces. *Political Geography, 30*(7), 370–380. https://doi.org/10.1016/j.polgeo.2011.08.001

Swyngedouw, E. (2016). The mirage of the sustainable "smart" city. Planetary urbanization and the spectre of combined and uneven apocalypse. In O. Nel-lo & R. Mele (Eds.), *Cities in the 21st century* (pp. 134–143). London: Routledge.

Taylor Buck, N., & While, A. (2017). Competitive urbanism and the limits to smart city innovation: The UK future cities initiative. *Urban Studies, 54*(2), 501–519.

Thatcher, J., O'Sullivan, D., & Mahmoudi, D. (2016). Data colonialism through accumulation by dispossession: New metaphors for daily data. *Environment and Planning D: Society and Space, 34*(6), 990–1006. https://doi.org/10.1177/0263775816633195

Till, K., & McArdle, R. (2016). The improvisional city: Valuing urbanity beyond the chimera of permanence. *Irish Geography, 48*(1), 37–68. https://doi.org/10.2014/igj.v48i1.525

Townsend, A. (2013). Smart cities: Buggy and brittle. *Places Journal.* https://doi.org/10.22269/131007

Trencher, G. (2019). Towards the smart city 2.0: Empirical evidence of using smartness as a tool for tackling social challenges. *Technological Forecasting and Social Change, 142*, 117–128. https://doi.org/10.1016/j.techfore.2018.07.033

Turner, D. (2016). *Digital denied: The impact of systemic racial discrimination on home-internet adoption.* Free Press. Retrieved from http://www.freepress.net/sites/default/files/resources/digital_denied_free_press_report_december_2016.pdf

van Dijk, J. (2006). *The network society: Social aspects of new media* (2nd ed.). Thousand Oaks, CA: Sage Publications.

Vanolo, A. (2014). Smartmentality: The smart city as disciplinary strategy. *Urban Studies, 51*(5), 883–899. https://doi.org/10.1177/0042098013494427

Vanolo, A. (2016). Is there anybody out there? The place and role of citizens in tomorrow's smart cities. *Futures, 82*, 26–36. https://doi.org/10.1016/j.futures.2016.05.010

Vanolo, A. (2018). Cities and the politics of gamification. *Cities, 74*, 320–326. https://doi.org/10.1016/j.cities.2017.12.021

Voytenko, Y., McCormick, K., Evans, J., & Schliwa, G. (2016). Urban living labs for sustainability and low carbon cities in Europe: Towards a research agenda. *Journal of Cleaner Production, 123*, 45–54. https://doi.org/10.1016/j.jclepro.2015.08.053

Wachowski, L., & Wachowski, L. (1999). *The matrix.* Retrieved from www.imdb.com/title/tt0133093/

Wagner, B. (2018). Ethics as an escape from regulation: From "Ethics-Washing" to ethics-shopping? In E. Bayamlioğlu, I. Baraliuc, L. Janssens, & M. Hildebrandt (Eds.), *Being profiled: Cogitas ergo sum: 10 years of profiling the European citizen* (pp. 84–89). Amsterdam: Amsterdam University Press. https://doi.org/10.2307/j.ctvhrd092.18

Wallace, R., & Wallace, D. (1980). Communications – Rand-HUD fire models. *Management Science, 26*(4), 418–422. https://doi.org/10.1287/mnsc.26.4.418

Ward, K. (2010). Towards a relational comparative approach to the study of cities. *Progress in Human Geography, 34*(4), 471–487. https://doi.org/10.1177/0309132509350239

Willis, K. S., & Aurigi, A. (2017). *Digital and smart cities.* New York and London: Routledge.

Yeung, K., Howes, A., & Pogrebna, G. (2019). *AI governance by human rights-centred design, deliberation and oversight: An end to ethics washing.* SSRN Scholarly Paper ID 3435011. Rochester, NY: Social Science Research Network. https://doi.org/10.2139/ssrn.3435011

Zuboff, S. (2019). *The age of surveillance capitalism: The fight for the future at the new frontier of power.* New York: Public Affairs.

Index